THE PHYSICAL TREATISES OF PASCAL

NUMBER XXVIII OF THE
RECORDS OF CIVILIZATION
SOURCES AND STUDIES

AUSTIN P. EVANS, *Editor*

RECORDS OF CIVILIZATION

SOURCES AND STUDIES

EDITED UNDER THE AUSPICES OF THE

DEPARTMENT OF HISTORY, COLUMBIA UNIVERSITY

THE PHYSICAL TREATISES OF PASCAL

THE
EQUILIBRIUM OF LIQUIDS
AND
THE WEIGHT OF THE MASS
OF THE AIR

Translated by
I. H. B. and A. G. H. SPIERS

with introduction and notes
by
FREDERICK BARRY

OCTAGON BOOKS

A DIVISION OF FARRAR, STRAUS AND GIROUX

New York 1973

Reprinted 1973
by special arrangement with Columbia University Press

OCTAGON BOOKS
A Division of Farrar, Straus & Giroux, Inc.
19 Union Square West
New York, N. Y. 10003

Library of Congress Catalog Card Number: 75-147407
ISBN 0-374-90418-9

Printed in U.S.A. by
NOBLE OFFSET PRINTERS, INC.
New York, N.Y. 10003

FOREWORD

THE *Traitez de l'equilibre des liqueurs et de la pesanteur de la masse de l'air* of Pascal, a modest little volume which summarizes completely his own brief but brilliantly ingenious labors in natural science, together with the remarkably well executed investigations of Perier which completed them, was put together by Perier and published at Paris in 1663, a year after Pascal's death. In its original form it was twice reprinted, first in 1664 and again in 1698; and it is included both in the comprehensive *Œuvres de Pascal* edited by Brunschvicg and Boutroux (14 volumes, 1904-14) and in the later *Œuvres complètes* of Strowski (3 volumes, 1923-31).

The original work and its early reprints are all now somewhat scarce, and the two editions of the *Œuvres* are usually accessible only in the larger libraries. For this reason it has been thought well worth the effort to prepare the following translation of this scientific classic. It is one of the few that provide the student of the history of science with a succinct yet entertainingly vivid picture of the spontaneous enthusiasm and experimental ingenuity that characterized the scientific work of the seventeenth century; and because it logically develops a single thesis, it exhibits, as no commentary could, the powerfully synthetic thought of this master of scientific method. To him we owe the first conclusive proof of the pressure of the atmosphere and the final banishment from the minds of natural philosophers of that ancient and persistent conception of *horror vacui* which had long inhibited investigation in this field; a complete mechanical correlation of all the diverse phenomena of fluid equilibrium, without recourse to that or any other imaginary conception; and finally, the establishment of detailed analogies between the effects of pressure in liquids and in air which affected the unification of hydrostatics and aerostatics as one deductively organized and logically coherent discipline.

To the modern reader, these treatises are peculiarly interesting as striking early illustrations of our own preferred scientific method of hypothesis and verification. Pascal, who had seen the Torricellian experiment but was probably not fully informed concerning Torricelli's reasons for his positive conviction that the seeming suspension of mercury in the barometric tube was due to atmospheric pressure, appears to have been led to his researches by a realization which the Italian physicist seems not fully to have grasped: that this explanation provided the means not only for a remarkable clarification of understanding but for the more significant elimination from natural philosophy of one of those plausible but imaginary conceptions, inherited from antiquity, which the more acute among his mechanically minded contemporaries already recognized as merely verbal generalizations, themselves inexplicable, but still operative psychologically to retard the progress of inductive knowledge in all fields of inquiry. This is probably not too much to assume in the case of a man whom we know to have possessed a remarkably penetrating logical mind. At all events, he seized upon Torricelli's explanation *ex hypothesi*, and with a remarkably fertile constructive imagination which was directed by a habit of thought essentially mathematical deduced its consequences with regard to a multiplicity of possible physical situations, perceived with the utmost clearness the analogy between this suppositional atmospheric pressure and the hydrostatic pressure which had recently been explained by Stevin, interpreted both with reference to the same conceptual imagery, and as his thought developed devised the most ingenious of experiments to verify his primary deductions with respect to each and to establish the similarity of the two types of effects by parallel tests conducted in like manner.

It is this reciprocal interplay of deductive and experimental procedure which especially stimulates one's interest in Pascal's work. In the end, the verification of all his contentions is complete; and if his manner of expression frequently suggests (and at times makes it evident, as Boyle pointed out) that certain of his tests were imagined rather than performed, this defect is not characteristic, but only occasional; and it may be ascribed, plausibly enough, to the impatience of a man who was conscious of the adequacy of experimental demonstrations actually carried out to prove his contentions, and to whom the physical performance of others, worth citing for the sake of convincing argument, might well have seemed an act of supererogation. The careful experimentalist today is keenly aware that such presupposition, however reasonable it may appear, is frequently very hazardous; but those who are less familiar than he is with the insidious influences of unsuspected superimposed effects will almost certainly consider Boyle's criticism to have been over-meticulous; and this is evidence enough that in Pascal's time it was no great fault, especially in a man whose predominantly deductive habit of thought was exceptionally cogent, that he did not mistrust the power of reason to that extent. It was this power of reason, after all, which achieved the most significant result of his research, for this was the theoretical correlation of a wide diversity of phenomena; and it distinctly adds to his deserved fame as a scientist that unlike more than one of his mathematical contemporaries, he was careful, sufficiently if not invariably, to substantiate his inferences by appeal to fact.

Aside from their value as exemplifications of scientific method, the several experimental records which this book contains possess another in their occasional references to the related investigations of other men. It has seemed that some of these demand a word or two of comment, though the greater number do not require it; and a few brief notes having once been written, it has been easy to yield to the temptation to amplify them a little, in order to recall to memory the wider range of investigation of which Pascal's work was a part. These notes, together with the translations of excerpts from the mentioned writings of Stevin, Galileo and Torricelli which have been placed in the Appendices, will, it is hoped, add enough to the historical value of the little classic to justify their inclusion.

It is a pleasure to acknowledge here, with the sincerest thanks, our obligation to Mrs. Barry and to Dr. Cioffari for the translations from Stevin's *Hydrostatique* and from the correspondence of Torricelli, which make fairly complete this first of the scientific series in the *Records of Civilization.*

FREDERICK BARRY

NEW YORK,
March 1, 1937

PREFACE

Containing the reasons which have called for the publication of these two Treatises after the death of M. Pascal, and an account of the various experiments which are explained therein.

ALTHOUGH several intelligent people who have read these two Treatises have expressed very favorable opinions about them, and although there are to be found in them a large number of the most marvelous effects in nature explained not by means of uncertain conjecture but by clear, sensible and demonstrable reasons, it may nevertheless be said with truth that the name of M. Pascal does greater honor to these works than they do to the name of M. Pascal.

Not but that these Treatises are so perfect of their kind that they could hardly be improved upon; it is the kind itself which is so greatly beneath him that those who judge of him only by these writings will be able to form but a very slight and very inadequate idea of the greatness of his genius and the quality of his intellect.

For although he had as great a gift as any man who ever lived for penetrating into the secrets of nature, concerning which he had admirable insight, he had so thoroughly realized for more than ten years before his death the vanity and the emptiness of all such knowledge, and had conceived such a distaste for it, that he could hardly suffer people of intellect to make it their study or the subject of their serious conversation.

From that time forth it was his firm belief that religion

was the one worthy object of the thoughts of men, and that it was one of the proofs of the degradation into which they had fallen through sin that they could devote themselves ardently to those things which cannot contribute to their happiness. He would frequently say on this subject *that all the sciences could not comfort them in the days of affliction, but that the doctrines of Christian truth would comfort them at all times both in affliction and in their ignorance of these sciences.*

Thus he believed that, though it might be of some advantage and an obligation of a sort to follow the general practice of inquiring into these subjects and ascertaining what might be said most reasonably and soundly concerning them, it was absolutely necessary to learn not to overrate them: and if it were better to know them without setting much store by them than not to know them at all, it would be better not to know them at all than to know them and overrate them and devote oneself to them as though they were great and lofty studies.

For this reason, although to the knowledge of several people who saw them in those early days these two Treatises were ready for the press more than twelve years ago, he nevertheless refused to allow them to be published, so loath was he always to put himself forward and so contemptuous of the value of these sciences.[1]

But it is not to be wondered at that his friends, who find themselves robbed by his death of the hope of several very

[1] This pitiful picture of a broken spirit will be sympathetically viewed only by those who realize from a study of Pascal's life the poignancy of his physical suffering and the extreme religiosity of the influences that surrounded him; it will be fully understood by none save those to whom the burden of life has at some time become, like his own, insupportable excepting for the consolations of faith. To explain the apparent sympathy of his biographer with this revulsion of feeling it is not necessary to recall the fears which were implanted in the minds of his French contemporaries by the fate of Galileo; his sympathy is probably quite genuine and exaggerated only by the promptings of a devoted loyalty. Consult in this connection F. Strowski, *Pascal et son temps* (1907-8); Boutroux, *Pascal* (1900); G. Michaut, *Les Époques de la pensée de Pascal* (1902); or in English, Viscount St. Cyres, *Pascal* (1910). The relevant psychology is thoroughly discussed by William James in *The Varieties of Religious Experience* (1912), especially in Lectures VI to X.

important works to which he intended wholly to devote himself in the service of the Church, should take a different view of the few writings which he has left them and therefore should be more eager to give them to the public.

For, deploring as they do the loss they have suffered, they cherish all that is left them of him. It revives the memory of one that was passing dear to them for many reasons, and continually provides them suggestions of that inimitable eloquence with which he spoke and wrote on all subjects that were susceptible of it. It is true that their intimate understanding of Monsieur Pascal's intelligence enabled them to discover in these fragments several things which will escape the notice of those who did not know him as they did. But it is believed also that all intellectual people will notice in them a most unusual skill in setting things in their true light, and will easily discern that the extraordinary lucidity which is apparent in his works is due to the remarkable clearness of his intellectual grasp.

And if they look a little further, and try to imagine what could have been produced by so admirable an insight and penetration, supplemented by so marvelous an abundance of rare and weighty thoughts and of vivid and striking expressions, when he devoted himself no longer to trivial speculations like those of these two Treatises but to the highest and loftiest truths of our religion, they will be able to form some conception of what M. Pascal could have done, had he been spared a greater length of years, in the works he had proposed to write and of which he has left only the slight beginnings; albeit these will not fail of admiration if they are ever given to the public.

The use that should be made of those that meanwhile are presented here is this: that they should not be valued for themselves nor allowed by their content to determine any estimate of their author; they should be looked upon as the sport and amusement of his youth and as things by which he set as little store as any man, and should serve only as an indication of what might be looked for from him in the serious and important studies to which he resolved to devote the rest of his life.

It is with the same object in view that I feel impelled to say something about his great aptitude for mathematics and of the way in which he learned it. It is perhaps a rarer and a stranger thing than has ever been heard of and should greatly contribute to the understanding of the quality of his intellect.

Monsieur Pascal never had any teacher but his father. The latter thought that he could make no better use of the leisure which he enjoyed after resigning the presidency of the Court of Aids at Clermont than to direct, himself, the studies of his son, whose aptitude for learning seemed to him to give the greatest promise. Such was the main reason which induced him to leave a provincial town in order to take up his residence in Paris where he thought his plans could be more easily carried out. He noticed particularly in the boy an admirable quickness in penetrating to the roots of things and in discriminating between sound arguments and those that were merely verbal. When one of the latter sort was suggested to him, his mind could not rest satisfied but remained in constant agitation until he had discovered true reasons for it. On one occasion (when he was still but eleven years of age), someone at table having inadvertently struck a dish of faïence porcelain with a knife, he noticed that it gave out a loud sound, which, however, immediately ceased when a hand was laid upon it. He at once tried to ascertain the reason of this; and the experiment having led him to make many more on the nature of sound, he discovered so much about it that he wrote a little treatise on the subject which was declared to be very ingenious and very sound.

This strange inclination towards matters of reasoning justly led his father (who was one of the best mathematicians in France) to fear lest, should he be allowed any acquaintance with geometry, he might delve too deeply into it to the neglect of the study of languages. The father resolved, therefore, so far as possible to keep all knowledge of it from him, locked up all the books that treated of it, and indeed refrained from mentioning it among friends in his presence. These precautions, however, merely excited the curiosity of the son, who often begged his father to teach him mathematics and, failing this, at least to tell him what that science was. President Pascal answered evasively that it was a science which taught how to draw correct figures and to ascertain the relations that they bore to one another. At the same time he forbade the boy to talk about it or indeed to think about it further. But this was a command that a boy with such a mind could not obey. Accordingly on that mere hint he fell to dreaming about it in his hours of relaxation, and when alone in his playroom he would take a piece of charcoal and draw figures on the tiles of the floor and seek the ways of making, for instance, a perfectly round circle, or a triangle with equal sides and angles, or the like. All this came to him easily; and then he would seek to ascertain the relations of these figures to one another. But since his father had taken such pains to hide all these things from him, he did not know even their names and was obliged to invent his own definitions. He called a circle a "round," a line a "bar," and so forth. After these definitions he made himself some axioms [and so advanced] until at last he made perfect demonstrations; and, as geometry is a progressive science, he proceeded so far in his investigations that he reached the thirty-second proposition of the first book of Euclid.

One day when he was thus engaged, his father happened into the room where he was working and found him so absorbed that for a long time he did not notice the intrusion. It is hard to tell which was the more surprised, the son to see his father who had expressly forbidden him this study, or the father to see his son in this maze of figures. But the father's surprise was far greater still when, having asked the boy what he was doing, the latter said that he was trying to find out something, which happened to be that very thirty-second proposition of the first book of Euclid. When asked what had put this into his head, the boy answered that it was due to his having found such or such another thing; and thus, retracing step after step and using all the time the words "bar" and "round," he finally got back to the definitions and axioms which he had invented for himself.

The elder Pascal was so appalled by the greatness and the power of his son's genius that he walked away unable to utter a single word and went forthwith to see his close friend, M. Le Pailleur, who was also very proficient in mathematics. When he reached him he stood perfectly still like a man in a trance. M. Le Pailleur, seeing this and noticing that he was even shedding tears, was greatly amazed and besought him no longer to withhold the cause of his distress. "I weep not for sorrow," said M. Pascal, "but for joy. You know the great pains that I have taken to keep geometry a sealed book to my son lest he be diverted from his other studies. And now see what he has done!" And thereupon he told him the whole story that I have just related and set forth in detail all that the boy had discovered for himself. M. Le Pailleur was no less surprised than the father and told him that he did not think it right to hold any longer such a mind in bondage and to conceal these sciences from him; that he should allow him access to the books that treated of them without further restriction. M. Pascal yielded to this judgment and gave the elements of Euclid to his son, who was still only twelve years of age. Never did a boy read a storybook more greedily and more easily than young Pascal read this book as soon as it was put into his hands. He perused it and understood it by himself without ever needing any explanation, and he went at once so deeply into the study that from that time forth he attended regular weekly meetings where all the cleverest people in Paris used to meet together, bringing their own works and examining those of others. The young Pascal had his place there at once with the best of them, both as regards examination and production. His contribution of novelties was as frequent as anybody's, and at times it happened that he discovered errors in the propositions under examination which had escaped the notice of the others. Still he devoted to the study of geometry only his leisure hours, the rest of the time being engaged in learning the languages that his father taught him. But as he found in these sciences that truth to which he was so passionately devoted, he made such progress, however brief the time he gave them, that at the age of sixteen he wrote a treatise on conic sections which the best mathematicians pronounced to be one of the greatest imaginable efforts of the human mind. Indeed M. Descartes, who had long been a resident of Holland, having read it and having heard that it had been written by a boy of sixteen, chose to believe that M. Pascal Senior was really its author but that he preferred to rob himself of the glory which was properly his in order to hand it on to his son. He could not believe that a lad of that age could possibly write a work of such power. By refusing to believe a thing that was in reality a fact, he showed how incredible and prodigious he thought it.

At the age of nineteen Pascal invented that admirable arithmetical machine which is considered one of the most extraordinary things that have ever been seen. And later, at the age of twenty-three, having witnessed Torricelli's experiment, he devised and made a large number of new ones. Since the two Treatises on the Equilibrium of Liquids and on the Weight of the Air deal with these experiments, we must here give a more precise history of them and go back a certain way.

History of the Experiments with a Vacuum

IT WAS Galileo who first noticed that suction pumps could not raise water higher than thirty-two or thirty-three feet and that any part of the tube which was higher than that remained apparently vacuous. He had merely inferred from this that nature abhors a vacuum only up to a certain limit and that the effort she makes to avoid it is finite and may be overcome; but he did not detect the falsity of the fundamental principle itself.[2] Later, in 1643, Torricelli, mathematician to the Duke of Florence and successor to Galileo, made the following experiment. He took a glass tube four feet long, open at one end and sealed at the other, filled it with quicksilver, stopped the open end with the finger or otherwise, and set it up vertically with the stoppered opening down a distance of two or three finger widths below the level of quicksilver, in a dish filled half with quicksilver and half with water. When the stopper was removed, the open end of the tube being still in the quicksilver of the dish, the quicksilver in the tube fell part way down and left in the upper part a space apparently empty, while the lower part remained full of the same quicksilver up to a certain height. When, now, the tube was raised somewhat until its open end, which formerly dipped in the quicksilver of the dish, emerged from the quicksilver and reached the zone of the water, the quicksilver in the tube rose to the top with the water, and the two liquids intermingled until at last all the quicksilver fell to the bottom and the tube was left full of water.

Such is the first experiment of the kind ever made: since then, it has become so famous by the inferences derived from it that it has always been called the Experiment with a Vacuum. It was the Rev. Father Mersenne, of the Order of Minims in Paris, who first heard of it in France. The news was sent to him from Italy in 1644, and he in turn spread it abroad and made the experiment famous throughout the country to the admiration of all scientists. M. Pascal learnt it from M. Petit, chief of the Department of Fortification, a very able scientist who had got it from Father Mersenne himself. They reproduced it together at Rouen in 1646 just as it had been performed in Italy, duplicating in every particular what had been reported from that country.

Later M. Pascal, having several times repeated the same experiment and established its accuracy, drew from it several inferences for the verification of which he made a number of

[2] See p. 73, n. 5.

new experiments in the presence of the most important men of the town of Rouen where he was then residing, his father being at the time chief of the Departments of Justice and of Finance. Among others he made one with a glass tube forty-six feet long, open at one end and hermetically sealed at the other, which he filled with water or rather, for greater visibility, with red wine. He had it set up vertically with the open end stopped and immersed in a dish of water to the depth of about one foot. On withdrawing the stopper the wine in the tube dropped to a height of some thirty-two feet above the surface of the water in the vessel, at which level it hung, leaving at the top of the tube thirteen feet apparently empty. When the tube was inclined, since the height of the wine contained in it would be thereby lessened, its level rose until it reached the height of thirty-two feet, and at last when the slope was increased so that the total height was reduced to thirty-two feet, it filled the tube entirely, water having replaced the wine previously expelled. Thus the tube became visibly full of wine from the top to a level thirteen feet from the bottom and full of water in the thirteen feet below, because water is heavier than wine.

He made, further, a large number of all sorts of experiments with syphons, syringes, bellows, and various tubes of different lengths, sizes, and shapes, filling them with various liquids such as quicksilver, water, wine, oil, air, and so forth.

In the year 1647 he had descriptions of these experiments printed in a little pamphlet[3] which he sent all over France and then abroad to Sweden, Holland, Poland, Germany, Italy, in every direction; which made them famous in the whole scientific world of Europe.

In that same year 1647 M. Pascal heard of an idea that had occurred to Torricelli, that air had weight and that its weight might be the cause of all the effects which had hitherto

[3] Entitled *Expériences nouvelles touchant le vuide.* This is referred to in the following treatises as *l'abbregé.* It is reprinted in Brunschvicg and Boutroux, and in Strowski, *op. cit.,* II, 53-76, and I, 11-21, respectively.

been ascribed to the abhorrence of a vacuum. This thought seemed to him wholly admirable, but as it was pure conjecture without anything approaching a proof,[4] he made several experiments in order to ascertain its truth or falsity. One of the most important of these was that of a vacuum within a vacuum which he made with two tubes one inside the other toward the end of the year 1647, as may be gathered from what is said about it in the account of the experiment at Puy de Dôme, which was printed in 1648.[5] No mention was made of it in the two Treatises that we are now publishing because its result is in every particular similar to that of the experiment related in the Treatise on the Weight of the Air, Chapter 6.[6] They are alike save for the fact that in the one there is used but one pipe while in the other there are two, one inside the other.

But still somewhat unsatisfied by this experiment, he conceived, at about the end of that same year, 1647, the celebrated experiment which was carried out in 1648 on the summit and at the base of a mountain in Auvergne called the Puy de Dôme, the account of which he printed and circulated far and wide.[7]

[4] An injustice to Torricelli, which appears to be inexcusable unless it is taken to imply that, so far as Pascal was aware at the time he heard about this idea, it had not been confirmed by experiment. As a matter of fact, Torricelli had not only inferred from the variability of the barometric height with changing conditions that the pressure which sustained his column of mercury was external, but had shown by experiment that it was independent of the volume of the space above this column within the tube, and therefore independent of the degree of rarefaction of any subtle matter from whatever source that might be assumed to exist in that space. See Appendix III, pages 163 ff.

[5] See note on this experiment, p. 100, n. 5.

[6] Page 56.

[7] Here it would appear that Perier, who probably wrote this account, does not a little injustice to himself—either through self-effacement, out of an exaggerated deference to Pascal, or through honest lack of realization that it was the admirable method of checks which, so far as the evidence indicates, he himself devised and applied that made this experiment conclusive and justly famous. When requesting him to carry out the experiment, Pascal wrote: "Ie ne touche pas aux moyens de l'executer, parce que je sçay bien que vous n'obmettrez aucune des circonstances necessaires pour la faire avec precision." See p. 102, below. The account here referred to is Perier's own, which appears below, pp. 103 ff.

The success of that experiment, which he later duplicated[8] several times on the top and at the base of various towers such as those of Notre Dame in Paris, of St. Jacques de la Boucherie, and others in the attic and cellar of a house, every time recording the same proportion,[9] wholly confirmed his belief in the theory of Torricelli concerning the weight of the atmosphere and led him subsequently to derive from it several very beautiful and useful consequences and to make many other experiments. These he described in a great Treatise which he wrote at that time, in which he explained very fully the whole matter and answered all the objections with which he had been met. But this Treatise was lost; or rather, since he was a great advocate of brevity, he himself condensed it into the two little treatises which are here published, one of them entitled "On the Equilibrium of Liquids" and the other "On the Weight of the Mass of the Air."

Of that other longer manuscript there have remained only a few fragments. These will be found at the back of this book; and together with them the narrative of the Puy de Dôme experiment of which mention has just been made.

It was immediately after this that more serious studies, to which M. Pascal wholly devoted himself, so put him out of conceit with mathematics and physics that he gave them up entirely. This is true despite the fact that he later wrote a treatise on the roulette,[10] under the name of Ettonville; for upon the subject matter of this work he merely happened accidentally without the slightest study; and he wrote it only to serve a purpose wholly foreign to mathematics and to all the sciences of physical investigation, as will some day be made clear. But, although from the year 1647 until his death almost fifteen years elapsed, it may still be stated that he actively

[8] See below, p. 109.

[9] Proportionality between difference of altitude and difference of barometric height.

[10] The cycloid. For Pascal's work on the cycloid see Brunschvicg et Boutroux, *Œuvres de Pascal,* VII, VIII, and IX, scattered chronologically; or, Strowski, *Œuvres complètes,* I, 233-56, 351-61.

lived but a very short time afterwards. His illnesses and his continual distresses scarcely left him a respite of two or three years during this time; and these were years not of perfect health, for he never enjoyed that, but only of a more supportable ill-health which did not wholly debar him from work.

It was in that short space of time that he wrote all that he has left behind him, both what appeared under other names and what has been found among his papers. This consists almost exclusively of a mass of detached memoranda intended for a great work that he had in mind, and he produced them in the brief intervals of leisure which his other occupations left him or in the conversations which he held with his friends. Nevertheless, although these thoughts are as nothing compared with what he would have done had he worked continuously and earnestly upon those books, it may be confidently asserted that if ever the public sees them it will be truly grateful to those who took the pains to collect and preserve them and will be convinced that these fragments, formless though they are, can hardly be overvalued, since they open great vistas of noble thought that no one else, perhaps, would ever have conceived.

[F. PERIER]

NOTICE

Due notice having been given that the first of the Tables of Figures to be found at the back of this book goes with the Treatise on the Equilibrium of Liquids and the second with the Treatise on the Weight of the Mass of the Air, two important remarks must be made, the first concerning the first Treatise and the other concerning the second.

First Remark

There is a statement in the Treatise on the Equilibrium of Liquids (page 5) which reads, "If the tube that was filled with water were a hundred times wider or a hundred times narrower, then so long as the level of the water in it remained the same, the same weight always would be required to balance it." This should not be accepted without the qualification that these tubes must always be above a certain size, say, two or three lines[11] in diameter. For if there were two communicating tubes of which one is no larger than perhaps the thickness of a pin, the water would stay at a higher level in the finer than in the wider tube. Even when these very fine tubes are apart, if you thrust them in water you see that the water rises and remains at levels some higher and some lower, according as the tubes are more or less fine, though they may be open at the top as well as at the bottom. But M. Pascal could hardly make this exception to the rule, because when he wrote these two Treatises the new experiments with very fine tubes had not yet been devised. Their invention is due to M. Rho,[12] a scientist wonderfully skilled in the devising and explaining of experiments.

[11] That is, from 5 to 7 millimeters.

[12] Jacques Rohault (1620-75), the most notable expositor of the Cartesian physics, whose *Traité de physique,* first published in 1671, was then held in the

Second Remark

Wherever the word "vacuum"[13] is found, it must not be imagined that M. Pascal intended to prove that there could be an absolute void: by this word he always means a space empty of all such matters as are perceptible to the senses, as he specifically states in various places.

It must also be observed that there is an error in the figure on page 56: the part "B" is not curved enough: the quicksilver left there does not fill it entirely but leaves an empty space. Hence it would happen that when the finger was removed the entering air would not drive the quicksilver left in that place ahead of it because it would have a passageway through which it could freely enter and fill the tube.

[F. PERIER]

---

highest esteem. In this work (Pt. 1, ch. xxii, secs. 81-85, and preceding) Rohault correctly describes at some length a variety of capillary phenomena, including those shown by liquids in communicating tubes one of which is of small diameter. His interesting explanations of these effects are wrong, however. The phenomena were first measured and correctly explained by Hawksbee in his *Physico-Mechanical Experiments* (1709).

[13] *Vuide*. This was already common usage, Stevin having formally defined it in this manner half a century before (see Appendix I, p. 138).

EXTRACT FROM THE KING'S PRIVILEGE

BY LETTERS Patent from the King issued in Paris on the second of April, 1663, over the signature of Bousselin and sealed with the Great seal: permission is granted to M. Perier, His Majesty's Councillor in the Court of Aids of Clermont-Ferrand, to print, sell, and distribute throughout the Realm of His said Majesty, by whatsoever bookseller and printer he may please, the works of M. Pascal his brother-in-law, entitled "Treatises on the Equilibrium of Liquids and on the Weight of the Mass of the Air, with a Few Experiments and a Few Fragments on the Same Subject."

This privilege to extend over a period of seven years beginning on the day when the printing of said works is completed for the first time; and all persons, of whatever quality or condition, are prohibited from printing, selling, or distributing under any pretext whatsoever any part of the same without the consent of said M. Perier or those empowered by him, under penalty of the seizure of the illegal copies, presses, fonts, and other instruments used in the manufacture of the illegal copies; of all costs, damages, and interest, and a fine of three thousand livres payable without deduction by each of the illegal printers and payable as appears in said Letters.

And the said M. Perier has chosen Guillaume Desprez, bookseller in Paris, to sell and distribute said "Treatises on the Equilibrium of Liquids and on the Weight of the Mass of the Air, etc.," in accordance with the agreement passed between them.

Entered in the books of the guild of booksellers and printers in accordance with the decree of Parliament of April 8, 1653, on the eighth day of June, 1663.

Signed, J. DuBray, Syndic.

First printing completed on the twenty-seventh day of November, 1663.

The required copies have been duly supplied.

CONTENTS[1]

A TREATISE ON THE EQUILIBRIUM OF LIQUIDS

TREATISE ON THE WEIGHT OF THE MASS OF THE AIR

[1] The greater part of this Table of Contents, referring to the works of Pascal, is translated from the text, where it appears at the end of the volume, under the heading "Index."

FRAGMENT

ANOTHER FRAGMENT

TREATISES ON THE EQUILIBRIUM OF LIQUIDS

AND

ON THE WEIGHT OF THE MASS OF THE AIR

CONTAINING THE EXPLANATION OF THE CAUSES OF VARIOUS EFFECTS OF NATURE WHICH HAD NOT BEEN KNOWN HITHERTO, AND IN PARTICULAR OF THOSE WHICH HAD BEEN ASCRIBED TO THE ABHORRENCE OF A VACUUM

By M. PASCAL

PARIS. PUBLISHED BY GVILLAUME DESPREZ, RUE ST. JACQUES, AT THE SIGN OF ST. PROSPER. M.DC.LXIII. WITH ROYAL PRIVILEGE

A TREATISE ON THE EQUILIBRIUM OF LIQUIDS[1]

CHAPTER I

How the Weight of Liquids is in Proportion to their Height

IF ONE fastens to a wall several receptacles, one as in the first figure [of Plate I],* the next sloping as in the second figure, another very wide as in the third, still another narrow as in the fourth, and the last merely a fine tube which ends in a broader but very short base as in the fifth, and if one fills them all with water to the same level, makes apertures of the same area at the base of each, and puts stoppers in to prevent the water from leaking out, experiment shows that it takes the same force to keep those stoppers in, although there are very different amounts of water in the several receptacles. This is because the water stands at the same level in all of them, and the measure of that force is the weight of the water contained in the first tube, which is uniformly of the same diameter throughout. If that water weighs a hundred pounds it will take a force of one hundred pounds to hold up each of the stoppers, even that in the fifth tube; though the water in it may weigh no more than one ounce.[2]

* Plate I faces p. 20.

[1] *Traité de l'equilibre des liqueurs.*

[2] This had long before been demonstrated by Stevin and by Galileo. As a matter

To test this with accuracy it is necessary to close the base of the fifth tube with a round piece of wood wrapped with tow like the plunger of a pump which fits this opening so exactly that it can slide up and down without sticking and yet prevent the water from escaping: a thread must then be fastened to the center of the plunger, passed through the small pipe, and fastened to one arm of a pair of scales, on the other arm of which a one-hundred-pound weight is hung. The hundred-pound weight will be seen to be in perfect equilibrium with the ounce of water in this small tube; and however little this one-hundred-pound weight is diminished, the weight of the water will bring down the plunger and with it the arm of the scales to which it is attached, while the arm to which the weight of a little less than one hundred pounds is attached will be seen to rise.

If this water happens to freeze and the ice does not stick to the tube (which it seldom does) the other arm of the scale need only carry a one-ounce weight to balance the ice; but if you apply heat to the tube and thaw the ice, it will take a one-hundred-pound weight to balance the weight of the now melted ice, although, as we have assumed, the weight of the water is just one ounce.

The same thing happens when these stoppered apertures are at the side or even at the top of the broader base; indeed that arrangement would make the experiment easier. To show this, it is necessary to have a container [like the rectangular

---

of fact Pascal made no fundamentally novel discoveries in hydrostatics, the value of his work being dependent rather upon the remarkably ingenious confirmatory tests which he invented and carried out and, particularly, upon the general correlations he discerned and with singular conclusiveness demonstrated. His failure to give credit to Stevin for his discoveries may possibly have been due to ignorance of his work; for Stevin's liberality of thought, no less than his political activities, had so stirred up feeling against him among Catholics in the Low Countries, that after his death powerful efforts were made among scholars there to consign him, by a conspiracy of silence, to oblivion. On the other hand, it is far from unlikely that Pascal himself shared this prejudice. On this matter, see Hoefer's article on Stevin in the *Nouvelle biographie universelle.* For Stevin's work in hydrostatics see Appendix I, and for that of Galileo, Appendix II.

base in Figure VI, page 20][3] hermetically sealed on all sides, on the top of which two holes are made, the one being narrow, the other wider, to which pipes are soldered that fit exactly. If now a piston is placed in the larger pipe and water is poured into the smaller, a great weight must be put on the piston to prevent the weight of the water in the smaller pipe from driving it up, just as in the former cases a force of a hundred pounds was necessary to prevent the weight of the water from driving the stoppers down, the opening being in the base. Likewise in the present case, if the opening had been in the side, the same force would have been required to prevent the weight of the water from pushing the piston towards that side.

And if the tube that was filled with water were a hundred times larger or a hundred times smaller, then so long as the level of the water in it remained the same, the same weight always would be required to balance it; and if the weight were diminished by never so little the water would flow down and cause this lesser weight to rise.[4]

RULE FOR THE AMOUNT OF FORCE NECESSARY TO PREVENT THE WATER FROM FORCING THE STOPPER DOWN

But if the water were poured to twice the height in the tube, it would take twice the weight on the piston to balance the water, and, similarly, if the piston opening were doubled in area the force necessary to keep the piston up would have to be doubled. Whence it appears that the force necessary to prevent the water from escaping by an opening is proportional to the height of the water and not to its expanse; and that the measure of its force is always the weight of all the water contained in a column as high as the water and of the same area as the opening.

What I have said of water must be understood to apply to every kind of liquid.

[3] For this and the following figures (VI to XVII) see plate I, facing p. 20.
[4] Cf. p. xxi, First Remark.

CHAPTER II

Why Liquids Weigh[1] in Proportion to their Height

ALL these examples show that a fine thread of water can balance a heavy weight. It remains to demonstrate the cause of such multiplication of force. This we can do by the following experiment.

NEW TYPE OF MACHINE FOR MULTIPLYING FORCES

Let there be a vessel full of water [Fig. VII], sealed on all sides and provided with two apertures, one of them one hundred times as large as the other. If a perfectly fitting piston is adjusted to each, one man pressing on the smaller piston will exert a force equal to that of one hundred men pressing on the larger, and will exceed that of ninety-nine men doing the same. And whatever the relative areas of these apertures, if the forces exerted upon the pistons are in the same ratio as the areas, they will balance one another. Whence it follows that a vessel filled with water is a new mechanical principle and a new machine that will multiply forces to any amount desired; for a man by this means will be enabled to lift any weight that another may propose.

It is remarkable that this new machine exhibits the same constant relation that is characteristic of all the old machines, such as the lever, the wheel and axle,[2] the endless screw, and others, which is that the distance traversed increases in the same proportion as the force.[3] For it is obvious that since one of these apertures is one hundred times as large as the other, a man pressing the small piston down a distance of one inch would move the other piston up only one-hundredth of that distance. It is the continuity of the water between the pistons

[1] *Pesent:* here meaning "exert downward pressure."
[2] *Tour.*
[3] *Le chemin est augmenté en mesme proportion que la force.*

that makes it impossible to move one without moving the other. It is clear that if the small piston moves one inch, the water thus moved presses on the other piston, and since the aperture of this one is one hundred times larger, it rises to only one-hundredth of the height. Thus the distances traveled are in the same ratio as the forces.[4]

Such may even be taken as the true cause of this effect, since it is evident that it amounts to the same thing whether we make one hundred pounds of water move through one inch, or make one pound of water move through one hundred inches. Thus when one pound of water and one hundred pounds of water are so arranged that the hundred pounds cannot move one inch without moving the one pound a distance of one hundred inches, they must stand in equilibrium, since one pound has as much efficacy[5] to make the hundred pounds move through one inch, as the hundred pounds have to make one pound move through one hundred inches.

For greater clearness it may be added that the water is equally pressed upon under the two pistons; for though one of these is one hundred times as heavy as the other, it is, on the other hand, in contact with an area a hundred times greater. Consequently, the pressure on each is the same and they must stand unmoved; for there is no reason why the one should overbear the other. Thus if a vessel filled with water has but one aperture, say one inch in area, and a piston is placed on it under a one-pound weight, that weight is exerted on every part of the vessel because of the continuity and fluidity of water. To ascertain how much of the weight is borne by each part, the following rule holds good.

Each area equal to that of the aperture, that is, one square inch, bears the same pressure as if it were pressed by the one-pound weight. The weight of the water is not taken into consideration here since only the weight of the piston is being

[4] *Le chemin est au chemin, comme la force à la force.* A more careful expression of this relation is given below, pp. 8 f.

[5] *Force.*

dealt with.[6] The one-pound weight presses the piston at the aperture, and each part of the vessel bears more or less pressure in proportion to its area, whether that part be opposite the aperture, or to one side of it, or far, or near; for the continuity and fluidity of the water make all those circumstances equal and indifferent.

The substance of which the vessel is made must be tough enough to withstand such pressures at every point. Should the resistance be inadequate at any point, then the vessel will burst; if the vessel is stronger than necessary, it furnishes the required resistance together with an excess resistance which is not useful in the circumstances. Should a second aperture be made in the vessel, then to stop the jet of water issuing from it a force will be required equal to the resistance needed by that part of the vessel; that is to say, a force which shall be to that of one pound as the area of the second aperture is to the first.

Here is another proof which will appeal only to geometricians, and may be disregarded by others.

I assume as a principle that a body never moves by its own weight without lowering its center of gravity.[7] From this follows the proof that the two pistons shown in Figure VII are in equilibrium, as follows: Their common center of gravity is the point that divides the line joining their individual centers of gravity in the proportion of their weights. Now, supposing it were possible for them to move, they would move through distances inversely[8] proportional to their weights, as we have shown. And if we observe their common center of gravity in this new position, it will be found at precisely the same point

[6] Pascal here corrects an error of oversight in the earlier corresponding treatment of the equilibrium of fluids in communicating vessels, by Galileo. See Ernst Mach, *The Science of Mechanics,* 4th English ed. (Chicago, 1919), pp. 90-91.

[7] The Principle of Torricelli, enunciated in his *De motu gravium naturaliter descendentium et projectorum,* published in his *Opera geometrica* (1644). See *Opere di Evangelista Torricelli,* edited by Loria and Vassura (1919) II, 105. The relevant passages are translated into French and discussed by Pierre Duhem: *Les Origines de la statique* (1905-6), II, 2 ff.

[8] *Reciproquement.*

as before: for it is always at the point which divides the line joining their individual centers of gravity in proportion to their weights. Therefore, owing to the parallelism of their lines of movement it will be at the intersection of the two lines joining the centers of gravity in each of the two positions. Hence the common center of gravity will be in the same place as before; hence the two pistons, considered as one system, have moved without lowering their common center of gravity, which violates the principle; hence they cannot move; hence they are at rest, that is to say, in equilibrium—which was to be proved.

I have demonstrated by this method, in a little Treatise of Mechanics, the reason of all the multiplications of force to be found in all the other mechanical devices invented so far. For I show that in all of them unequal weights, when made to balance with the aid of machines, are so disposed by the machinery that their common center of gravity is not lowered, whatever their position; whence it follows that they must remain at rest, that is, in equilibrium. Let us then accept it as beyond doubt that if a vessel filled with water and provided with two apertures undergoes pressures at these apertures in proportion to their areas, these forces will be in equilibrium. Such is the foundation and reason of the Equilibrium of Liquids, of which a few examples follow.

THIS NEW MECHANICAL MACHINE EXPLAINS WHY LIQUIDS WEIGH IN ACCORDANCE WITH THEIR DEPTHS

This mechanical machine for multiplying forces, if rightly understood, shows why liquids weigh[9] in accordance with their depth and not in accordance with their areas in all the effects that we have set down. For in Figure VI it is evident to the eye that the water in a small tube balances a piston bearing a weight of one hundred pounds, because the vessel at the base is a vessel full of water, with two apertures, over one of which is a large piston, while over the other is the column of

[9] *Pesent,* (v. *supra*).

water which is virtually a piston with a weight of its own. These two must balance each other if their weights are to one another as the areas of the apertures. Again, in Figure V [page 20] the water in the small tube is in equilibrium with a one-hundred-pound weight. The reason is that the vessel at the base, which is wide but shallow, is a hermetically sealed vessel filled with water, with two apertures, one wide, at the bottom where the piston is, and the other narrow, at the top where the small tube is. The water in the latter is virtually in itself a piston and balances the other because of the proportionality of the weights to the areas of the apertures; for, as we have said, it does not matter whether the apertures face each other or not.

It is obvious that the water in these tubes behaves exactly as would copper [or brass][10] pistons of the same weights, since a brass piston weighing one ounce would balance the one-hundred-pound weight just as well as does the thin thread of water weighing one ounce. Thus the cause of the equilibrium between a light weight and a heavier, as seen in all these instances, is not that these light weights which balance much heavier weights consist of liquids (for this is not the case in every instance, as the same results are obtained when small copper pistons balance such heavy ones) but is that the matter which fills the bases of the vessels, from one aperture to the other, is liquid. This is a feature common to all the instances given and is the true cause of the multiplication. Further, if in the Figure V the water in the slender tube were to freeze and that in the tube with a wide base were to remain liquid, it would take one hundred pounds to hold up the weight of that ice; but if the water in the base were to freeze, whether the rest froze or not, one ounce would suffice to balance it.

From this it is most clearly evident that it is the liquidity[11]

[10] *Cuivre.* Boyle translates this as "brass" (p. 21, n. 1), not impossibly, because the probability here seems to be that pistons would be made rather of brass than of copper, which is less readily worked; but below (p. 15) the specific gravity of *cuivre* is given as 9, which indicates copper.

[11] *Liquidité* (mobility).

of the substance by which one aperture communicates with the other that causes this multiplication of force. The fundamental reason is, as we have said, that a vessel full of water is a mechanical machine for multiplying force.

Let us now proceed to the consideration of other effects of which this machine reveals the cause.

Chapter III

Illustrations[1] *of the Equilibrium of Liquids*

If a vessel filled with water has two apertures [Fig. VIII], to each of which a [cylindrical] tube is soldered, and if water is poured into each tube up to the same level, the two columns will be in equilibrium. Their heights being the same, their volumes will be proportional to their cross sections:[2] that is, proportional to the areas of their apertures. Thus these two columns of water are in reality two pistons whose weights are in the ratio of the aperture areas. The foregoing demonstrations, therefore, show them to be in equilibrium. Hence if water is poured into one of these tubes it will force the water up in the other until the same level is reached in both; when the two will be in equilibrium, for then they will be a pair of pistons whose weights will be in the same ratio as their aperture areas. This is the reason why water rises to the height of its source.

WHY WATER RISES TO THE HEIGHT OF ITS SOURCE. THIS KIND OF EXPERIMENT CAN BE PERFORMED ONLY BY FILLING THE BASE-VESSEL WITH THE HEAVIER OF THE LIQUIDS TO THE MOUTHS OF THE TUBES

If different liquids are poured into the tubes, say, water in one and quicksilver in the other, the two liquids will be in

[1] *Exemples et raisons.*

[2] The actual statement is looser: *elles seront en la proportion de leurs grosseurs.*

equilibrium when their heights are in the same ratio as their weights. In this case the water will be fourteen times as high as the quicksilver, since quicksilver is fourteen times as dense as water,[3] and we shall have two pistons, one of water and the other of quicksilver, the weights of which will be proportional to their aperture areas. And even if the water-filled tube were one hundred times more slender than that filled with the quicksilver, that fine thread of water would balance the whole great mass of quicksilver, provided it were fourteen times as high.

All that has been said thus far of tubes must be taken to apply to any vessel, whether regular or irregular in shape: for the same equilibrium is established in all cases. Thus if, instead of the tubes which we have pictured at these two apertures, two other receptacles were adapted to them, wide in places and narrow elsewhere—in short, irregular throughout their lengths—then should liquids be poured into them to the heights specified above, these liquids will be as balanced in these irregular tubes as in the regular, because it is their height and not their width[4] that determines the pressure of liquids.

This could be readily demonstrated by inserting in each of the tubes several smaller regular tubes; for thus we could show, from the foregoing demonstrations, that any corresponding pair of these inserted tubes, one in the first of the larger tubes, the other in the second,[5] are in equilibrium; whence it would follow that all those of one large tube will be in equilibrium with all those in the other. Those who are familiar with inscribing and circumscribing in geometry will find no difficulty in understanding this; to others it would be very difficult to offer a geometrical demonstration.

If one sets up in a river a tube bent backward at its lower extremity and completely full of quicksilver, and always keeps

[3] *Parce que le vif argent pese de luy mesme quatorze fois plus . . .*
[4] *Largeur.*
[5] *Que deux de ces tuyaux inscripts, qui se correspondent dans les deux vaisseaux . . .*

its upper end above the surface of the water [Fig. IX], the quicksilver will drop part of the way down [the longer arm] until it reaches a certain level, after which it falls no more, but remains at a height which will be one-fourteenth of the height of the water above the recurved lower end. Thus if from the surface of the water to the recurved end the distance is fourteen feet, the quicksilver will drop until it reaches a height of one foot only above that end, and there it will stay: for the weight of the quicksilver inside will balance that of the water outside the tube, since their heights are in proportion to their weights, while their widths are immaterial to the equilibrium. Similarly it is also immaterial whether the recurved end is wide or narrow, and consequently sustains a large or a small weight of water.

Again, if the tube is sunk deeper in the water, the quicksilver rises, for the weight of water has increased. On the other hand, if the tube is raised, the quicksilver falls, for its weight is now greater than the other. If the tube is inclined, the quicksilver rises until it reaches the necessary level, which had been lowered by the inclining, since a tube when inclined is of lesser height than when it is vertical.

The same effect is produced with a simple, that is to say, a straight tube. Let the tube be open top and bottom, filled with quicksilver and set in a river [Fig. X]. So long as the upper end is out of the water, if the lower end reaches down to a depth of fourteen feet the quicksilver will sink to the level of one foot only [above the lower end] and will be held there by the pressure of the water. This is readily understood, for the water, pressing the quicksilver from below and not from above, strives to push it up, as though driving a piston, with the more force as it is greater in height, until, when the weight of the quicksilver [thus raised] exerts the same force downward that the water exerts upward, the whole system comes to equilibrium.

Further, it is obvious that, if there were no quicksilver in the tube, the water would enter it and would rise to a height of fourteen feet, this being the surface level. Therefore, since one foot of quicksilver weighs the same as the fourteen feet of water which it replaces, it naturally comes to the same equilibrium [with the water outside] that fourteen feet of water in the tube would establish.

If, however, the tube were lowered until the upper end was submerged, then the water would enter the tube and the quicksilver would drop out; for, since the water would then press downward within as well as without the tube, the quicksilver would lack the counterweight necessary to hold it up.

Chapter IV

On the Equilibrium between a Liquid and a Solid

WE WILL now give examples of equilibrium between water and solid bodies, as, for instance, that between water and a solid copper[1] cylinder.

The cylinder can be made to float in this way. Take a long tube, say one about twenty feet in length, which is enlarged at its lower end, funnelwise [Fig. XI]. Into the round mouth of that end insert a copper cylinder, so accurately turned that it can slide in and out of the mouth without allowing any water to escape. It is not difficult to construct such a piston. Now set up this whole system in a river, so that only the upper end emerges; support it with the hand and leave the cylinder free to move as it will. The solid cylinder, then, will not sink but will float, because it is in contact with the water beneath and not above, since no water can enter the tube. Thus the water presses it up just as it pressed the quicksilver up in the last experiment, and with just the same force as the weight of copper exerts to make the cylinder sink, so that the two op-

[1] Cf. notes on pages 10 and 15.

posing forces balance each other. Of course, to bring about this result the tube must be immersed deeply enough to give the water a head[2] sufficient to counterpoise the copper. If the cylinder is one foot in height, there must be nine feet from the surface of the water to the bottom of the cylinder, since copper by itself weighs nine times as much as water.[3] If the column of water is not long enough, as for instance, when the tube is lifted higher in the water, the weight of the cylinder prevails and it sinks. But if the tube is lowered too deep, say to a depth of twenty feet, then it certainly will not sink by its own weight; on the contrary, a strong force will be necessary to wrench it apart from the funnel, since the weight of the water drives it upward with the force of a head of twenty feet. But if a hole is bored in the tube and the water, thus let in, presses upon the cylinder both above and below, then the cylinder sinks by its own weight, like the quicksilver in the other experiment, because it is no longer supported by a counterpoise.

If such a tube as we have just pictured is bent upward, a cylinder of wood put in it, and the whole system plunged in water with the upper end of the tube just emerging [Fig. XII], the wood will not float up, although it is surrounded by water. On the contrary, it will sink in the tube, because the water is in contact with it above but not below, since it cannot get inside the tube. Consequently the water presses the cylinder downward with all its weight and exerts no upward pressure because it is not in contact with the cylinder below.

Now if the cylinder were made to float low,[4] just so as to have no water above it, without emerging at all [Fig. XIII], it would bear no pressure of water either above or below, since the water, unable to enter the tube, would be in contact with it neither above nor below. Only at the sides would there be contact all round, and the cylinder would not rise, since noth-

[2] *Hauteur.*

[3] This specific gravity indicates that *cuivre* here (and probably elsewhere) means not brass, but copper.

[4] *Que si ce Cilindre estoit à fleur d'eau . . .*

ing would press it upward; on the contrary, it would sink by its own weight only.

And if the lower end of the tube were twisted sideways, like a crutch, and a cylinder inserted, and the whole system immersed in water with the upper end emerging, the weight of the water would drive the cylinder across and into the tube, because the water would not be in contact with it on the opposite side and would therefore exert a force that would be the stronger, the greater its depth in the water.

Chapter V

On Bodies Wholly Immersed in Water

WE HAVE seen that water presses upward bodies that it bears upon from below, that it presses downward those that it bears upon from above, and that it presses to one side those that it bears upon from the opposite side. From this it can be readily inferred that, when a body is wholly submerged [Fig. XV], then, since the water bears upon it above, below, and on every side, it strives to push it up, down, and to all sides; but as its head is the measure of its force in all these efforts, there is no difficulty in determining which of them should overbear the rest. It is obvious at once that since the water has the same height on all the lateral faces, it will press upon them equally; and the immersed body, consequently, will receive no particular impulsion towards any side, any more than would a weather vane between two equal winds. But as the water has a greater head against the bottom than against the top, it will obviously press the body more upward than downward; and since the difference between these heights of water is the height of the body itself, it will be readily understood that the water presses it upward and not downward, and with a force equal to the weight of a volume of water equal to that of the body.

A BODY IN WATER IS BALANCED BY AN EQUAL VOLUME OF WATER

Thus a body in water is buoyed up by the same force that would lift it if it were in one of the trays of a pair of scales, while the other tray was weighted with a volume of water equal to its own.

THAT IS WHY SOME BODIES SINK

Therefore if it is of copper or other material that is heavier than water, volume for volume, it sinks, because its weight overbears that of the counterpoise.

OTHERS RISE

But if it is of wood or other material lighter than water, volume for volume, it rises with all the force by which the weight of the water exceeds its own.

OTHERS NEITHER RISE NOR SINK

And if the weights are equal, the body neither sinks nor rises. For instance, wax remains in water approximately where it is put in.

It follows that the bucket of a well is easy to raise so long as it is in the water: one does not feel its weight until it begins to emerge. Similarly a bucket filled with wax would also be easy to lift so long as it remained in the water. Not that water, or wax, does not weigh as much in the water as out of it; but in the water they have a counterpoise which they lack when they are taken out of it. Thus in the water they are easy to lift, just as the scale of a balance loaded with a one-hundred-pound weight is easy to lift if there is an equal weight in the other scale.

COPPER WEIGHS MORE IN THE AIR THAN IN WATER

Hence when copper is under water, it is found lighter by exactly the weight of an equal volume of water.[1] If it weighs

[1] The Principle of Archimedes. For the original demonstration see T. L.

nine pounds in the air, it weighs only eight in water, because it is counterpoised by an equal volume of water, weighing one pound. In sea water it weighs less, since sea water is about one-forty-fifth heavier.

TWO BODIES, IN EQUILIBRIUM IN THE AIR, ARE NOT SO IN WATER

For the same reason two bodies, one of copper and the other of lead, of equal weight and consequently of different volume (since it takes more of the copper to make the same weight) will be found in equilibrium if they are put in the two trays of a pair of scales. But if the scales are held under water, the equilibrium is destroyed: for each body being balanced by a volume of water equal to its own, and the volume of the copper being greater than the volume of the lead, the copper has a larger counterpoise and therefore is overborne by the lead.

NOT EVEN IN A DAMP ATMOSPHERE

Similarly, when two weights of different materials are balanced with the greatest accuracy which man can attain and are in perfect equilibrium in a very dry atmosphere, they cease to be so balanced when the atmosphere is humid.

WATER LIFTS IMMERSED BODIES BY ITS WEIGHT AND DOES NOT BEAR THEM DOWN

It is in accordance with the same principle that when a man is immersed in water, then, far from being borne down by the weight of the water, he is, on the contrary, lifted. But being heavier than water he sinks nevertheless, though with far less violence than in the air; for he is counterpoised by a volume of water equal to his own, weighing almost as much as himself. If it actually weighed as much, he would float. Hence if he gives himself an impetus by kicking bottom, or

---

Heath: *The Works of Archimedes* (1897), pp. 257-59. For Stevin's demonstration see Appendix I, Proposition VIII, pp. 144 f.

makes the slightest effort to oppose the water, he rises and floats. In mud baths, a man cannot sink; and if he is pressed down, he rises again of his own accord.

For the same reason, in a bathtub it is easy to lift an arm that is immersed; but out of the water the arm is felt to be quite heavy, because it lacks the counterpoise afforded by a volume of water equal to its own, which helped to raise it when immersed.

WHAT MAKES BODIES FLOAT

Lastly, floating bodies weigh just the same as the water which they displace: the water in contact with them below and not above, presses them upward only.[2]

That is why a convex plate[3] of lead will float: its shape causes it to displace a large volume of water. If it were in a lump it would never take up in the water a larger space than a volume of water equal to the volume of its own material, and that volume would be insufficient to counterpoise it.

Chapter VI

On Immersed Compressible Bodies

ALL that I have set forth shows the way water acts upon all bodies put into it; by pressing them on all sides. From this it is readily to be inferred that if a compressible body is immersed, it must be compressed inwardly toward the center. That is exactly what does happen, as the following examples will show [Fig. XIV].

If a bellows with a very long nozzle, say twenty feet in length, is immersed so that only the nozzle-tip emerges, then, provided the small air inlet-valves are plugged, it will be hard

[2] Archimedes's Law of Floating Bodies. For the original demonstration see T. L. Heath: *Works of Archimedes* (1897), pp. 255-57. For Stevin's demonstration, see Appendix I, Proposition V, pp. 142 f.

[3] *Platine*—here, clearly, one in the shape of a shallow dish.

to open, although in the air it could easily be opened: the water compresses it on all sides by its weight. If the necessary force is exerted, it may be opened; but the smallest relaxation of that force will cause it to slam shut again owing to the weight of water pressing on it: in the air it would remain open. And the deeper it is in the water, the harder it is to open, because there is a greater head of water to overcome.

For like reasons, if a twenty-foot tube is inserted into the air pipe of a balloon and bound securely [Fig. XVI], and if through this tube quicksilver is poured until the balloon is exactly filled, then when the whole is plunged into a tank of water until the free end of the tube just emerges, the quicksilver will be seen to rise from the balloon to a certain level in the tube. The reason is that the water presses the balloon on all sides; the quicksilver which it contains is similarly pressed at every point, save at the entrance of the tube, to which the water has no access, since the tube reaches above the surface. Thus the quicksilver is driven out of the parts where it is under pressure to the part where it is not, and so rises in the tube till it reaches such a height that its weight is the same as that of the water outside. The effect would be similar if the balloon were pressed between both hands, for the quicksilver could then be easily driven up the tube. This clearly shows that [when it is immersed] the surrounding water presses upon it in the same way.

It is for the same reason that if a man sets one end of a glass tube twenty feet in length upon his thigh, and then sits down in a tank full of water with the upper end of the tube just emerging [Fig. XVII], his flesh will rise under the mouth of the tube and a big and painful swelling will be formed there, as if his flesh were sucked up as it is in the process of cupping. The weight of the water compresses his body everywhere save at the mouth of the tube, where the water cannot reach since it is kept away by the walls of the tube. The flesh is driven from where it is compressed to the spot where it is not compressed. The higher the water-level, the greater

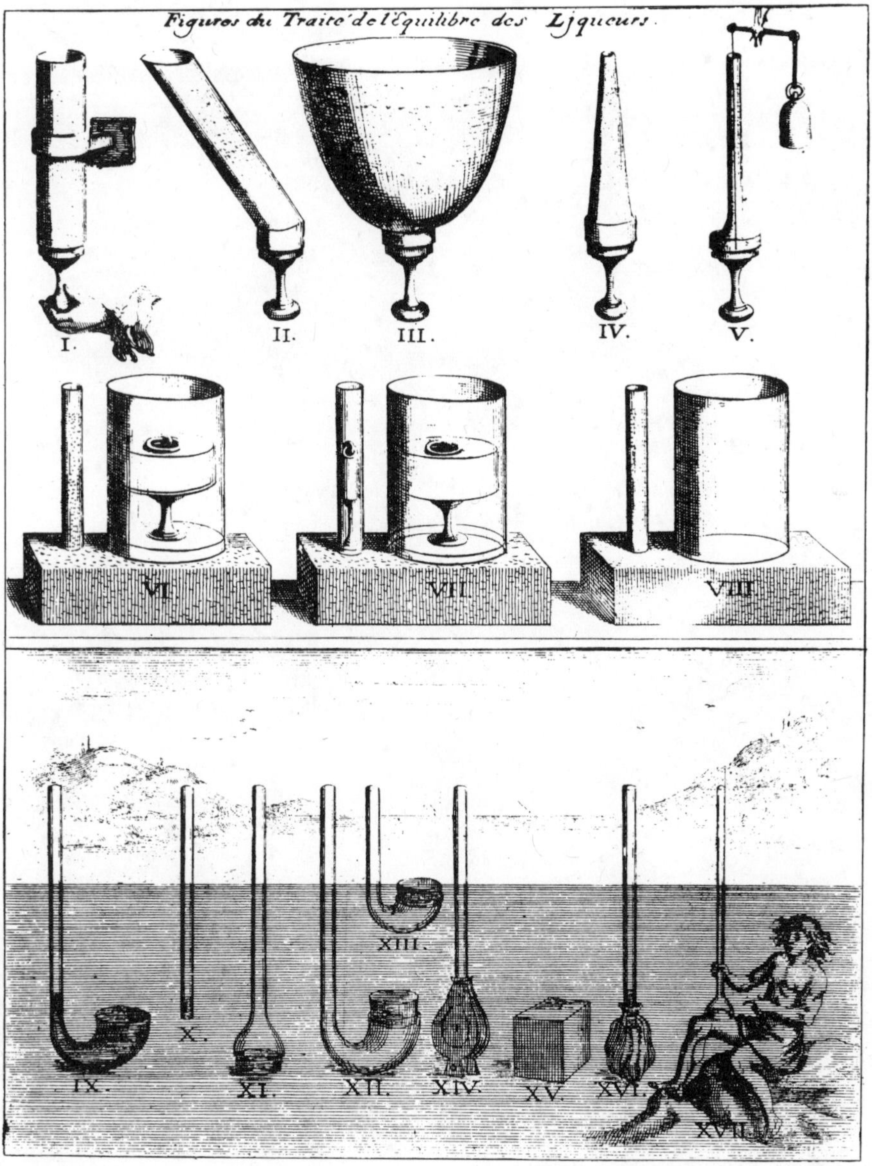

Figures du Traité de l'Equilibre des Liqueurs.

the swelling. On withdrawing the water the swelling disappears, just as it does if water is poured down the tube; for as the weight of the water then bears upon that spot as well as everywhere else, there is no more swelling there than elsewhere.[1] This result is exactly in accord with the preceding. The quicksilver in the one case, and the human flesh in the other, are pressed on all sides save just at the spots covered by the mouths of the tubes; and they are thus driven up the tubes as far as the weight of the water can lift them.

If one places at the bottom of a water-filled tub a balloon in which the air is only moderately compressed, it will show marked diminution of volume; and as the water is drained away it will gradually swell up. This is because the weight of the mass of water above the balloon presses it together on all sides towards the center until the resilience of the air thus compressed equals the weight of the incumbent water.

[1] It was this "experiment," particularly, that Boyle so amusingly satirized in his *Hydrostatical Paradoxes* (1666). As a good Baconian he insisted that all such statements should be justified by actual experience; a contention which Pascal, like many another logician of his day and later, would doubtless have considered the expression of a wholly superfluous caution. Boyle wrote, in part:

"And the reasons, why, notwithstanding that I like most of Monsieur *Pascal's* assertions, I decline employing his way of proving them, are principally these:

"FIRST, Because though the experiments he mentions be delivered in such a manner as is usual in mentioning matters of fact; yet I remember not that he expressly says that he actually tried them, and therefore he might possibly have set them down as things that *must* happen, upon a just confidence that he was not mistaken in his ratiocinations. And of the reasonableness of this doubt of mine I shall ere long have occasion to give an instance.

"SECONDLY, Whether or no Monsieur *Pascal* ever made these experiments himself, he does not seem to have been very desirous that others should make them after him. For he supposes the phaenomena he builds upon to be produced fifteen or twenty foot under water. And one of them requires that a man should sit there with the end of a tube leaning upon his thigh; but he neither teaches us how a man shall be enabled to continue under water, nor how, in a great cistern full of water twenty foot deep, the experimenter shall be able to discern the alterations that happen to mercury and other bodies at the bottom.

"AND thirdly, These experiments require not only tubes twenty foot long, and a great vessel of at least as many feet in depth, which will not in this country be easily procured; but they require brass cylinders, or plugs, made with an exactness that, though easily supposed by a mathematician, will scarce be found obtainable from a tradesman."—*Hydrostatical Paradoxes:* introductory letter to Lord Brouncker, in the Birch edition of the *Works*, 1772. Vol. II, pp. 745-46.

If, however, a balloon filled with very highly compressed air is placed at the bottom of the same water-filled tub, no compression will be visible. This is not because the water is not pressing on it; for, in the case of the other balloon, in that of the balloon with quicksilver, in that of the bellows, and in all the other instances given, the contrary fact was proved. The reason is that the water has not force enough to compress this air perceptibly, on account of its great previous compression: just as a very stiff spring, such as an archery bow, cannot be bent by a moderate force which would very visibly bend one of weaker resistance.

And it is not to be wondered at that the weight of the water does not perceptibly compress this balloon, although the mere application of a finger, of much less force, can compress it markedly. The reason for this difference of behavior is that when the balloon is immersed, it is pressed by water on all sides, whereas when it is pressed by a finger, it is pressed in one spot only. Now when it is pressed by a finger in one spot only, a deep dent can easily be made, because the adjoining parts are not pressed and thus can easily take up the air which is driven from the spot that is pressed. Thus, since the matter which is driven from the one spot pressed distributes itself through the rest, each spot has but little to find room for, and a very visible difference appears between the pressed part and all the other parts about it which bear no pressure. But if the other parts were to be pressed like the first, each part, giving up what it had received from the first, would come back to its former state, inasmuch as it would bear a pressure equal to that borne by the first. The result, therefore, would be nothing more than a general compression of all the parts toward the center, and no compression would be apparent in any particular spot. This general compression could be gauged only by comparing the space finally occupied by the balloon with that which it occupied originally; and since these volumes would be only very slightly different, it would be impossible to detect the change. This makes it apparent that pressure at one point only and general pressure on every part are very different.

The same thing happens when every part of a human body is pressed excepting one; for at that part a swelling develops which is caused by a pressure transmitted[2] from all the rest, as was seen in the case of a man immersed in water with a tube on his thigh.

Again, if the same balloon is squeezed between both hands, try as you may to cover every part, there will always be some that slip out between the fingers; and just at these points big swellings will appear. But, if it could be squeezed everywhere with the same pressure, it could never be perceptibly compressed by any effort, provided that the air in the balloon had previously been compressed vigorously. That is precisely what happens when it is in water, which touches it on every side.

CHAPTER VII

On Animals in Water

FROM all this we discover why water does not compress the animals within it, although it presses uniformly all the bodies it surrounds, as has been shown by many examples. It is not that the water does not press them, but that, as we have already said, it touches them on every side and therefore cannot cause either swelling or depression at any particular spot but only a general condensation of all parts toward the center. This condensation is imperceptible unless it is very great; and it must necessarily be very slight, owing to the compactness of the bodies of the animals. For if the water was only in partial contact with them, or everywhere except in just one place, provided its height was great, the effect

[2] *Regorgement.*

would be very noticeable, as we have seen; but if the water presses it everywhere, no change is visible.

WHY ONE DOES NOT FEEL THE PRESSURE OF WATER

It is easy to proceed from this point to the reason why animals in water do not feel its weight.

The pain we feel under pressure is great if the compression is great, because the part pressed has its blood squeezed out, and the flesh, nerves, and other parts of which it is composed are pressed out of their normal positions: a violence which cannot be unattended with pain. But if there is only slight compression as when the skin is so lightly touched with a finger that the blood is not squeezed away, nor the flesh or the nerves displaced, nor any similar change brought about, there can be no sense of pain. Whatever part of the body may thus be touched, we can feel no pain from so slight a compression.

That is exactly what happens to animals in water. The weight compresses them indeed, but not visibly, for the reason set forth above. No part is pressed, or squeezed bloodless, or disturbed in nerve, vein, or flesh; for everything being under the same pressure, there is no reason why anything should be pressed one way or another; and there being no change, there can nowhere be pain or indeed special sensation. It is, therefore, not to be wondered at that animals do not feel the weight of water, although they would distinctly feel the pressure of merely a finger, a pressure less than that of water. The reason for this difference is that when immersed, they are pressed on all sides uniformly, whereas when they are pressed with a finger they are acted upon at one point only. We have shown that this difference is why they are visibly compressed by a finger-tip touch, but not visibly at all by the weight of water, though this weight may be a hundred times greater. And as sensation is always in proportion to compression, that same difference is the reason why they feel easily enough a compression by a finger, and not at all the weight of the water. Thus the real reason why animals in water do not feel its weight is that they are pressed equally on all parts.

Similarly if a worm were put into a mass of dough, then although it were squeezed between the hands it could never be crushed nor even injured nor distorted, because it would be pressed on all sides. The following experiment will prove this. Into a glass tube, closed at the bottom and half-filled with water, drop three things—namely, a small balloon half-filled with air, another quite full of air, and a fly (which can live in luke-warm water as well as in the air); and then push into this tube a piston that reaches the water. If now you press upon the piston with whatever force you please, as, for instance, by piling many weights upon it, the water will press all that it contains; the half-filled balloon will be very noticeably compressed; but the taut balloon will be no more compressed than if it were under no pressure at all, nor will the fly; which will feel no pain under this heavy weight but will move freely and briskly along the glass, and if released from its prison, will fly off immediately.

It is not necessary to possess great clarity of thought to confirm by this experiment all we have thus far demonstrated. We see the weight pressing upon all these bodies with all its force; compressing the slack balloon, and consequently the taut one close to it, since the same reasoning applies to both, though the latter does not show any compression. Whence arises, whence must arise, this difference, if not from the single condition by which the two balloons differ, which is that one of them is full of air which has been forcibly introduced under pressure, while the other is only half full? The slack air in the latter can be greatly compressed, but that in the other cannot, because it is compact already, and the water surrounding it and pressing it on all sides can make no observable impression upon it, because it arches over it like a completely enveloping vault. The fly is not compressed. Why not, unless for the same reason that the taut balloon is not compressed? And finally the fly suffers no noticeable pain, for the same reason. If at the bottom of the tube dough were substituted for water, and both the balloon and the fly were thrust into it, the pressure caused by the piston above would produce the very same effect.

It follows, therefore, that the sole condition of totally surrounding pressure causes the compression to be neither painful nor perceptible. Must it not be granted, therefore, that this is the only reason why animals in water are insensible to the pressure upon them? Let us, then, no longer assign as the reason for this that water in water has no weight[1]: for it weighs everywhere alike; nor that it weighs otherwise than do solids: for all weights are alike, and we have just seen that a fly can support a solid weight without feeling it.

Do you wish for something yet more conclusive? Let the piston be removed, and let the tube be filled with water until the latter, which thus displaces the piston, weighs as much as the piston itself. Undoubtedly the fly will not feel the weight of this water any more than he felt that of the piston. Whence comes this insensibility to so heavy a weight in these two cases? Is it that the weight is water? No, for the same happens if the weight is a solid. Let us say then that the reason is solely that the animal is surrounded with water. That feature alone both cases have in common: therefore, it is the real reason.

Further, if it so happened that all the water above this animal froze, then so long as there remained just enough liquid above it as to surround it on every side, again it would not feel the weight of the ice any more than previously it felt the weight of the water. And if all the water in the river were to freeze down to within one foot from the bottom, the fish swimming in it would not feel the weight of the ice any more than that of the water into which it would subsequently melt. Thus animals in water do not feel its weight; not because it is merely water that weighs upon them, but because they are surrounded with water.

[1] *Que l'eau ne pese pas sur elle mesme.* An interesting indication of the survival among educated men of a conception long since disproved. Stevin's work had invalidated it half a century before this time—though perhaps, it must be admitted, without Pascal's knowledge. See n. 2, pp. 3 f.

TREATISE ON THE WEIGHT OF THE MASS OF THE AIR[1]

CHAPTER I

The Mass of the Air has Weight, And with this Weight presses upon all the Bodies it surrounds

IT IS no longer open to discussion that the air has weight. It is common knowledge that a balloon is heavier when inflated than when empty, which is proof enough. For if the air were light,[2] the more the balloon was inflated, the lighter the whole would be, since there would be more air in it. But since, on the contrary, when more air is put in, the whole becomes heavier, it follows that each part has a weight of its own, and consequently that the air has weight.

Whoever wishes for more elaborate proofs can find them in the writings of those who have devoted special treatises to the subject.[3]

[1] *Traité de la pesanteur de la masse de l'air.* See p. 62, n. 8.

[2] *Leger:* specifically light; that is, tending in all circumstances to move upward against gravity—an Aristotelian conception, applied by him to the element fire only (De Caelo 311) but by later Peripatetics applied also to air.

[3] The famous Milanese mathematician and philosopher of nature, Girolamo Cardano (1501-76), appears to have been the first to demonstrate that air has weight and to attempt a determination of its specific gravity. After him, and before it was known how to produce a vacuum, many experiments were made

If it be objected that air is light when pure, but that the air that surrounds us is not pure, being mixed with vapor and impurities which alone give it weight, my answer is brief: I am not acquainted with "pure" air, and believe that it might be very difficult to find it. But throughout this treatise I am referring solely to the air such as we breathe, regardless of its component elements. Whether it be compound or simple, that is the body which I call the air, and which I declare to have weight. This cannot be denied, and I require nothing more for my further proof.

This principle being laid down, I will now proceed to draw from it certain consequences.

1. Since every part of the air has weight, it follows that the whole mass of the air, that is to say, the whole sphere of

---

to establish his principal conclusion as indubitable fact. The best of these were carried out by Galileo and described in his *Discorsi e dimonstrazioni mathematiche* (1638), in the dialogue of the First Day. With a syringe he forced into a properly valved bottle, previously weighed, an amount of air which at ordinary pressure would have been of two or three times its volume, and determined with precision an actual increase of weight; then allowed the excess air to escape without loss through a tube which led into another bottle completely filled with water and so constructed that the water displaced by the entering air could be caught in a suitable weighed vessel, the increase in the weight of which, finally, he determined. Galileo remarked that if air were specifically light the first flask when filled with condensed air should be lighter than before, not heavier; that air consequently had weight (as, indeed, previous experience had already proved); and that since the volume of the water displaced was that of the excess air introduced, in its normal condition, the ratio of the determined weight of this air to that of the displaced water was the specific gravity sought. In a second, more elegant experiment, he forced water, instead of air, into the valved flask first used and thus compressed the air within it; weighed it in this condition; then opened the valve, thus allowing to escape a volume of air which, in its normal condition, had previously occupied the space now filled by the water; and finally weighed again. In this case, the increase in weight after the water had been introduced was the weight of this water, the subsequent loss in weight being that of the same volume of air in its normal condition. See the English translation of the *Discorsi* by Crew and De Salvio: *Dialogues concerning Two New Sciences,* New York, 1914.

A simpler, approximate determination was described by Descartes in 1642 (see *Œuvres de Descartes,* edited by Adam and Tannery, Paris, 1897-1910, Vol. III, p. 483). A glass bulb with a neck drawn out to small diameter was weighed at ordinary temperature; then strongly heated so that much of its contained air was driven out, sealed without perceptible loss, cooled, and weighed again.

the air, has weight, and as the sphere of the air is not infinite in extent, but limited, neither is the weight of the whole mass of the air infinite.

2. The mass of the water of the sea presses with its weight that part of the earth which is beneath it; if it surrounded the whole earth instead of only a part, its weight would press upon the whole surface of the earth. In the same way, since the mass of the air covers the whole face of the earth, its weight presses upon the earth at every point.

3. Just as the bottom of a bucket containing water is pressed more heavily by the weight of the water when it is full than when it is half empty, and the more heavily the deeper the water is, similarly the high places of the earth, such as the summits of mountains, are less heavily pressed than the lowlands are by the weight of the mass of the air. This is because there is more air above the lowlands than above the mountain tops; for all the air along a mountain side presses upon the lowlands but not upon the summit, being above the one but below the other.

4. Bodies immersed in water are pressed on all sides by the weight of the water above them, as we have shown in the Treatise on The Equilibrium of Liquids. In the same way bodies in the air are pressed on all sides by the weight of the air above them.

5. Animals in water do not feel its weight; neither do we the weight of the air and for the same reason. Just as it would be a mistake to infer that, because we do not feel the weight of the water when immersed in it, water has no weight; so it would be a mistake to infer that air has no weight because we do not feel its pressure. We have shown the reason of this in the Treatise on The Equilibrium of Liquids.

6. If there were collected a great bulk of wool, say twenty or thirty fathoms high, this mass would be compressed by its own weight; the bottom layers would be far more compressed than the middle or top layers, because they are pressed by a greater quantity of wool. Similarly the mass of the air, which

is a compressible and heavy body like wool, is compressed by its own weight, and the air at the bottom, in the lowlands, is far more compressed than the higher layers on the mountain tops, because it bears a greater load of air.

7. In the case of that bulk of wool, if a handful of it were taken from the bottom layer, compressed as it is, and lifted, in the same state of compression, to the middle of the mass, it would expand of its own accord; for it would then be nearer the top and subjected there to the pressure of a smaller quantity of wool. Similarly if a body of air, as found here below in its natural state of compression, were by some device transferred to a mountain top, it would necessarily expand and come to the condition of the air around it on the mountain; for then it would bear a lesser weight of air than it did below. Hence if a balloon, only half inflated—not fully so, as they generally are—were carried up a mountain, it would necessarily be more inflated at the mountain top, and would expand in the degree to which it was less burdened.[4] The difference will be visible, provided the quantity of air along the mountain slope, from the pressure of which it is now relieved, has a weight great enough to cause a sensible effect.

There is so necessary a bond between these consequences and their principle that if the principle is true the consequences will be true also. Since, therefore, it is acknowledged that the air, reaching from the earth to the periphery of its sphere, has weight, all the conclusions we have inferred from this fact are equally correct.

But, however certain these conclusions may be deemed, it appears to me that all who accept them would nevertheless be eager to see this last consequence confirmed by experiment, because it involves all the others and indeed directly verifies the principle itself. There is no doubt that if a balloon such as we have described were seen to expand as it was lifted up, the conclusion could not be avoided that the expansion was

[4] *Et qu'il deuroit s'élargir à proportion de ce qu'il seroit moins chargé:* an actual statement—*ex hypothesi*—of Boyle's Law, established in the year of Pascal's death.

due to a pressure, which was greater below than above. Nothing else could cause that expansion, the more so as the mountains are colder than the lowlands. The compression of the air in the balloon could have no other cause than the weight of the mass of the air, since this air was taken in its actual condition at low altitudes and was uncompressed, the balloon being even limp and only half inflated. This would be proof positive that air has weight; that the mass of the air is heavy; that its weight presses all the bodies it contains; that its pressure is greater on the lowlands than on the highlands; that it compresses itself by its own weight, and is more highly compressed below than above. And, since in physical science experience is far more convincing than argument, I do not doubt that everyone will wish to see this reasoning confirmed by experiment. Moreover, should the experiment be performed, I should enjoy this advantage: that if no expansion of the balloon were observed even on the highest mountains, my conclusions, nevertheless, would not be invalidated; for I might then claim that the mountains were still not high enough to cause a perceptible difference. Whereas if a considerable and very marked change occurred, say of one-eighth or one-ninth in volume, the proof, to me, would be absolutely convincing, and there could remain no doubt as to the truth of all that I had asserted.

But I delay too long. It is time to say, in a word, that the trial has been made and with the following successful result.

AN EXPERIMENT MADE AT TWO HIGH PLACES, THE ONE ABOUT 500 FATHOMS HIGHER THAN THE OTHER

If one takes a balloon half-filled with air, shrunken and flabby, and carries it by a thread to the top of a mountain 500 fathoms high, it will expand of its own accord as it rises, until at the top it will be fully inflated as if more air had been blown into it. As it is brought down it will gradually shrink by the same degrees, until at the foot of the mountan it has resumed its former condition.

This experiment proves all that I have said of the mass of the air, with wholly convincing force; but it must be fully confirmed, since the whole of my discourse rests on this foundation. Meanwhile it remains to be pointed out only that the mass of the air weighs more or less at different times, according as it is more charged with vapor or more contracted by cold.

Let it then be set down, (1) that the mass of air has weight; (2) that its weight is limited; (3) that it is heavier at some times than at others; (4) that its weight is greater in some places than in others, as in [highlands and] lowlands; (5) that by its weight it presses all the bodies it surrounds, the more strongly when its weight is greater.

CHAPTER II

The Weight of the Mass of the Air produces all the Effects hitherto ascribed to the Abhorrence of a Vacuum

THIS chapter is divided into two parts: the first describes the principal effects which have been commonly ascribed to the abhorrence of a vacuum; the second shows that they are due to the weight of the air.

PART ONE

An account of the effects ascribed to the abhorrence of a vacuum

There are several effects which nature is said to produce by an abhorrence of a vacuum, of which the most striking are the following:

I. When all the apertures of a bellows are closed, it is hard to open. Any attempt to do this meets with a resistance as if its two sides were stuck together. And similarly the piston of a sealed syringe resists the effort to withdraw it, as though it adhered to the base of its case.

This resistance is commonly explained as an instance of nature's abhorrence of the vacuum which would be produced if the bellows could be opened: and this theory is supported by the fact that the resistance ceases as soon as an aperture is made by which air can enter to fill the bellows when it is opened.

II. Two polished surfaces laid one upon the other are difficult to separate and seem glued together. Thus a hat laid on a table is hard to jerk up. A piece of leather pressed against a paving stone and jerked up, will wrench out and lift the stone.

It is claimed that this adhesion is due to nature's abhorrence of the vacuum that would exist while the air was passing from the periphery to the center.

III. When a syringe is dipped in water and the piston is drawn back, the water follows it, and rises as if it adhered to the plunger. In the same way, in a suction pump, which is really but a long syringe, the water rises and follows its piston when this is drawn back, as if it adhered to it.

It is claimed that this rising of the water is due to nature's abhorrence of the vacuum which would be left when the piston is withdrawn, if water did not take its place, since air cannot enter. This explanation is supported by the fact that if slits are cut through which air can penetrate, the water no longer rises.

Similarly, if the nozzle of a bellows is thrust into water and the bellows is opened suddenly, water enters to fill it because no air can get in, especially if the vents in the sides are sealed.

Again, if a man sucks up water, the same cause produces the result; for the lungs act like a bellows of which the nozzle is the mouth.

Again, in breathing, the air is drawn in as a bellows draws in air, to fill its emptiness.

Again, if lighted tow is put in a saucer full of water and

covered with an inverted glass, then, as the fire dies out the water rises in the glass, because the air within it, rarefied by the fire, is condensed as it cools and draws the water up with it to fill the place which its condensation has left empty; just as the piston of a syringe draws the water up behind it.[1]

Again, in cupping, the flesh is drawn up into a swelling; because the air inside the cup, rarefied by the flame of the candle, is condensed as it cools when the flame dies, and draws up the flesh to fill the vacated space, as the water was drawn up in the foregoing instance.

IV. If a bottle, filled with water, is set upside down in a water-filled vessel, the water hangs in the bottle and does not fall out.

It is claimed that this retention of the water is due to nature's abhorrence of the vacuum which would be produced were the water to drop away while no air could enter to fill the vacancy left behind. This explanation is supported by the fact that if a vent is made through which the air can flow in, the water drops immediately.

The same test may be made with a tube, say ten feet in length, sealed at the top and open at the bottom. If the tube is filled with water and the open end [temporarily closed] is dipped into a vessel full of water, the water will be wholly retained in the tube, whereas it would run out at once if the top were opened.

The same thing may be done with a similar tube, sealed at the top but bent backward at the lower [open] end, without dipping it in a water-filled vessel as in the preceding case. For if the tube is filled with water, the water will be retained, whereas if the top were opened a jet of water would instantly and violently escape from the bent extremity.[2]

Lastly, the same thing may be done with a simple straight tube, provided it be very narrow at its lower end. If it is sealed

[1] In this case, as is implied, the tow is only partially submerged, and is not at once extinguished. At first, some of the warmed air would probably bubble out through the water and thus suggest this interpretation. The chemistry of combustion was not, at this time, understood.

[2] The instrument now called the siphon barometer.

at the top the water will be retained, whereas the water would rush out below if the upper end were opened. This is why a wine-filled cask does not release a single drop, though the spigot be open, until a vent is opened at the top.

V. If a tube bent in the shape of an inverted horseshoe (which is commonly called a siphon) is filled with water, and its legs are placed so as to dip in separate water-filled vessels, then, however small the difference of level between the two vessels may be, all the water from the higher vessel will rise up the leg immersed in it to the top of the siphon and will pass down by the other leg into the lower vessel in which that leg is immersed; so that, if a sufficiency of water be supplied to the higher vessel, the flow will be continuous.

It is claimed that this retention of the water is due to nature's abhorrence of the vacuum that would be left in the siphon if the water from these two legs were to fall from each into its own. So it does, if any vent is made at the top of the siphon, whereby the air can enter.

There are several other similar effects which I do not describe because they are all like those I have mentioned. In all of them the one outstanding cause is that all bodies in contact resist any effort to part them when the air cannot insinuate itself between them, whether that effort be due to their own weight, as in the cases where the water rises and is retained in spite of its weight, or to the force exerted to part them as in the earlier examples.

Such are the effects commonly ascribed to nature's abhorrence of a vacuum. We will now show that they are due to the weight of the air.

PART TWO

The weight of the mass of the air produces all the effects that are commonly ascribed to the abhorrence of a vacuum

If it has been well understood, from the Treatise on the Equilibrium of Liquids, how their weight presses all the bodies immersed in them, there will be no difficulty in understanding how the weight of the mass of the air bearing on all bodies produces on them the effects that might be ascribed to abhorrence of a vacuum: for they are quite alike, as we shall prove in each case.

I. The weight of the mass of the air causes the difficulty in opening a sealed bellows.[3]

To make it clear how the weight of the mass of air causes the resistance encountered in opening a bellows from which the air is excluded, I will point to a similar resistance due to the weight of water. It needs only to be remembered, as I said in the Equilibrium of Liquids, that if a bellows (Fig. XIV) with a tube twenty or more feet long is set in a tank full of water with the tip of the nozzle extending above the surface, it is hard to open; and that the greater the depth of water above it the harder it is to open. This is obviously due to the weight of the water above: for if there is no water there, it is easy to open. The more water you pour in, the greater is the resistance, which is always equal to the weight of the water sustained. The reason is that as the nozzle projects above the water, and therefore excludes it, the bellows cannot be opened without raising and holding up the whole mass of water. The water that is pushed aside in the act of opening cannot enter the bellows, is forced to find room elsewhere, and thus raises the water level—a process attended with some difficulty; whereas if the bellows were so perforated that water could get in, it could be freely opened and closed because the water could enter through the perforations as fast as room was made for it, and would not, therefore, be lifted. I do not think that anyone can be tempted to ascribe this resistance to the abhorrence of a vacuum. It is absolutely certain that it is due solely to the weight of water.

Now what we say about water must be taken to apply to any other liquid: for if the bellows is set in a vessel full of

[3] Illustrations of the apparatus used to exhibit the phenomena here referred to will be found on Plate II, facing p. 57.

wine, the same resistance to its opening will be experienced; likewise with milk, oil, quicksilver, and indeed with any liquid whatsoever. Thus it is a general rule and a necessary effect of the weight of liquids, that if a bellows is so immersed in any one of them that the liquid is excluded from its interior, the weight of the liquid above makes it impossible to open the bellows without overcoming a resistance due to the fact that it has to be lifted. Applying this general rule to air in particular, it follows as a certain consequence that when a bellows is so sealed as to exclude all air, the weight of the air above prevents its opening without overcoming some resistance: since it cannot be opened without lifting the whole mass of air. But as soon as a perforation is made in the bellows, it can be freely opened and closed, because now the act of opening no longer lifts the mass of the air. All this is completely analogous to the action of the bellows immersed in water.

Whence it is evident that the difficulty in opening a sealed bellows is but a particular case of the general rule that it is hard to open a bellows in any fluid[4] whatsoever which is prevented from entering it.

What we have said about this effect we will say of all the rest, but more briefly.

II. The weight of the mass of the air is the cause of the difficulty that one feels in separating two polished bodies in close contact.

To explain how the weight of the mass of the air causes the resistance felt when the attempt is made to separate two polished surfaces in close contact, I will give an example of a wholly similar resistance due to the weight of water, which will put it beyond doubt that the air causes this effect. Here again what was stated in the Equilibrium of Liquids must be recalled.

Let a copper cylinder carefully ground on a lathe be placed

[4] *Liqueur,* as also above; but here obviously, and as previously implied, equivalent to *fluide.*

in the mouth of a funnel made with equal care, until they fit so perfectly that the cylinder enters smoothly into the funnel without any leakage of water between them; and let this device be plunged into a tank full of water so that the stem of the funnel—which may be made twenty feet long if necessary—just emerges (Fig. XI). If, now, while the funnel is held in one hand, the cylinder, at a depth of fifteen feet in the water, is released and left to move as it will, not only will it remain in position, although it seems to be quite unsupported; but furthermore, there will be difficulty in withdrawing it from the funnel, although it is in no way attached to it. On the other hand it would drop violently of its own weight if it were only four feet below the surface of the water in the tank, and more violently still if it were entirely out of the water. The reason for this I have already made clear. It is that water is in contact with the cylinder below but not above (since the funnel prevents its contact with the upper surface); it presses the face it touches toward the face it does not touch, and thus drives the cylinder up against the funnel.

The same reasoning applies to any other liquid.[5] Consequently, when two polished surfaces are laid together, if the upper is held in the hand while the lower is left free, the latter must remain suspended, because it is in contact with the air below, but not with the air above, since there is no opening between the two plates, and the air, consequently, cannot reach the surfaces in contact. Whence it follows, as a necessary effect of the weight of all fluids, that the weight of the air must drive the lower body up and press it against the upper so strongly that a great resistance will be felt to the effort of separating them: an effect completely analogous to the effect of the weight of water.

Thus it is evident that the difficulty of separating two smooth bodies is but a particular case of the general rule that applies to the pressure of all fluids whatever when they are

[5] *Liqueur* = *fluide,* and similarly, further on.

in contact with one of the surfaces of a body, but not with the surface opposite.

III. The weight of the mass of the air is the cause of the rise of water in syringes and pumps.

To explain how the weight of the mass of the air makes water rise in pumps as the plunger is drawn back, I will explain an entirely similar effect of the weight of water, which will account for it perfectly.

If a syringe is provided with a long piston, say ten feet in length and hollow throughout, with a valve at its base opening downwards and not upwards, it cannot suck water nor any other liquid above the level of the liquid, because air can freely enter through the hollow of the piston. When the mouth of this syringe is plunged into a vessel full of quicksilver, and the whole apparatus is placed in a tank full of water so that the top of the piston shall just emerge, then if the piston is drawn up, the quicksilver will also rise behind it, as though adhering; though it would not rise at all if there were no water in the tank, because the air would then have free access to the body of the syringe through the hollow neck of the piston.

There is here no abhorrence of a vacuum, for even if the quicksilver did not rise to fill the space vacated by the piston, there would be no vacuum; since the air could enter freely. The sole cause is the mass of water which presses on the quicksilver in the vessel from every side except at the mouth of the syringe (which it is prevented from reaching by the body of the syringe and by the piston). This quicksilver, then, pressed on all sides but one, is forced by the weight of the water towards that one side as soon as the rising piston leaves it entrance room, and balances, within the syringe, the weight of the water outside. But if slits are made to admit water into the syringe, the quicksilver will cease to rise, because the water enters by them and is now in the same contact with the mouth of the syringe as with the other parts. Since its pressure then affects all parts equally, no liquid rises. All this has been clearly demonstrated in the Equilibrium of Liquids.

This illustration makes it clear how the weight of the water causes the quicksilver to rise. The same effect might be produced by the weight of sand. If all the water is removed from this vessel, and sand is poured in, in its place, the weight of the sand will cause the quicksilver to rise in the syringe because, just as the water did before, the sand now presses it on every side save that which is at the mouth of the syringe; and pressing it, compels it to rise. And if you bear on the sand with your hands, you will drive the quicksilver further up the syringe until it reaches such a height that it can balance the extra pressure.

The explanation of these effects makes it very easy to see why the weight of the air causes water to rise in common syringes to the height that the piston is drawn back. The air, in contact with the water of the vessel on every side except at the mouth of the syringe (from which it is excluded by the syringe itself and by its piston), pressing by its weight upon that water on every side except that single one, cannot fail to drive it up as far as the withdrawing piston makes room for it to enter, and to counterbalance, within the syringe, the weight of the air outside. It does so for the same reason that the quicksilver rose under the pressure of the water, and under that of the sand in the case that we have just given, and by the same necessity.

Thus it is obvious that the rising of water in syringes is but a particular case of the general rule that any liquid pressed in every one of its parts save one by the weight of another fluid is thereby driven towards that part against which no pressure is exerted.

IV. The weight of the mass of the air causes water to be retained in tubes sealed at the top.

To show how the weight of the air causes water to be retained in tubes sealed at the top, we shall now give an example of such suspension caused by the weight of water. This will be perfectly convincing.

First of all it may be stated that this effect is only a variant of the preceding one. Just as the weight of the air causes the water to rise in syringes and retains it there, as we have shown, this same weight of the air similarly retains the water in a tube. In order that this effect may be considered, like the others, with reference to another with which it can be closely compared, we will refer to a statement already made in the Equilibrium of Liquids. When a tube ten or more feet in length, which is bent backward at its lower end is filled with quicksilver and then placed in a tank full of water until only its top emerges (Fig. IX), the quicksilver remains partly suspended in the tube, at such a height that it can balance the weight of the water outside. This same suspension occurs in an unbent tube, simply open at the top and at the bottom, provided the upper end be out of water.

Now it is apparent that this suspension is not due to abhorrence of a vacuum, but simply to the fact that the water exerts its weight outside the tube, and not inside; that it is in contact with the quicksilver on one side and not on the other, and therefore, by its weight, keeps the quicksilver suspended at a given height. And if the tube be perforated so that water may be let in, instantly all the quicksilver drops; for then the water is in contact with it everywhere, exerts pressure upon it inside as well as outside the tube, and has now nothing to serve as a counterpoise. All this has been set forth in the Equilibrium of Liquids.

This being a necessary effect of the equilibrium of liquids, it naturally happens that when a tube, sealed at the top and bent backward at the bottom, is filled with water, the water will be held suspended. For the air pressing with its weight that part of the water that is in the bend but not pressing that which is in the tube (because of the sealing), the water in the tube must necessarily be held suspended within the tube to balance the weight of water outside; precisely as the weight of the water held the quicksilver in equilibrium in the case we have just examined. The same thing happens when the tube is not bent. For air being in contact with the water below but not above (since the tube is sealed there), it is inevitable that the weight of the air shall hold the water up: just as the water, in our last example, holds up the quicksilver, and presses upwards and supports a brass cylinder with which it is in contact below but not above. But if the seal is broken the water falls, for the air is then in contact with the water both below and above, and exerts its weight both within and without the tube.

This makes it clear that the holding of liquids in suspension by the pressure of the air with which they are in contact on one side but not on the other is a case of a general rule that liquids contained in any sort of tube immersed in another fluid whereby they are pressed on one side but not on the other are held in suspension by fluid equilibrium.

V. The weight of the mass of the air causes water to rise in siphons.

To make it apparent how the weight of the air causes water to rise in a siphon, we will show that the weight of water makes quicksilver rise in a siphon which is fully open at the top and to which, therefore, air has free access. From this it will be clearly seen how the weight of the air produces this effect. And we will do so thus:

Let a siphon with one leg some twelve inches long and the other thirteen be opened at the top and let a tube twenty feet in length be soldered hermetically to this opening. Then let it be filled with quicksilver and placed with its legs dipping into separate vessels also filled with quicksilver; and let the whole apparatus be set up in a water-filled tank to a depth of some fifteen or sixteen feet, the upper end of the open tube remaining out of water. Now if one of the vessels be never so little, say one inch, higher than the other, all the quicksilver in the higher vessel will rise to the top of the siphon and make its way by the other leg to the lower vessel in a continuous stream; and if more quicksilver is fed to the higher vessel the flow will be continuous. But if the siphon is punctured so that water may enter it, instantly the quicksilver will drop out of each leg into its vessel, and water will take its place.

This rising of the quicksilver is not due to the abhorrence of a vacuum, for the air has perfectly free access to the siphon. Again, if the water were removed from the tank, the quicksilver from each leg would drop into its separate vessel, and air would take its place through the now freely open tube. Thus manifestly it is the weight of the water that causes this rising, because it presses upon the quicksilver in the vessels and not upon that in the siphon. This weight forces the quicksilver to rise and to flow as it does. But no sooner is water admitted by a perforation than, pressing now inside as well as outside the siphon, it ceases to force the quicksilver upwards.

For the same reason that water must necessarily cause the quicksilver to rise in a siphon when it presses upon the vessels without any access to the inside of the siphon, similarly the weight of the air must cause the water to rise in common siphons: it exerts its weight upon the water held by the vessels into which their legs dip, without any access to the inside of the siphon; but if a perforation is made so as to provide that access, the water ceases to rise, and on the contrary drops into each vessel and is replaced by air which now exerts its weight both inside and outside the siphon.

It is apparent that this last effect is but a particular case of the general rule, and that if it is clearly understood why the weight of the water makes the quicksilver rise in the example we have given, it will be apparent likewise why the weight of the air raises the water in common siphons. Hence we must make very clear the reason why the weight of the water produces this effect, and explain why it is the higher vessel that empties itself into the lower and not the lower into the other.

To do so, it must be observed that the water weighs upon the quicksilver in each vessel, but not on the quicksilver inside the legs that dip into it. Thus the quicksilver in the vessels is pressed, by the weight of the water, up each leg of the siphon to its top, and would rise farther if this were possible, for the water in the tank is sixteen feet deep while the siphon is but one foot in height; and one foot of quicksilver is equal to only fourteen of water. So, obviously, the weight of the water drives the quicksilver to the top of each leg, with some power left over. Thus the quicksilver in each leg being driven up by the weight of the water, the pressures in the two legs oppose each other at the top, and the stronger must prevail. It will be easy to calculate which this will be. Since the water has a greater height by one inch above the lower vessel, it drives up the quicksilver in the longer leg with more power than in the other leg, by the margin of the power derived from one inch of height. From this it would seem at first glance that the result would be to drive the quicksilver from the longer leg into the shorter. But it must be taken into consideration that the weight of quicksilver in each leg opposes the effort of water to press it up. These two resistances are not equal. Since the quicksilver in the longer leg is the deeper, by one inch, it offers a resistance which is greater by the force derived from one inch of height. Hence the quicksilver in the longer leg is driven up by a force of water the greater by the height of one inch. But it is borne down by its own weight, that is by the excess weight of one inch of quicksilver. Now one inch of quicksilver weighs more than one inch of water; therefore the quicksilver in the shorter leg is driven up with the greater force and consequently must continue to rise so long as there is quicksilver in the vessel into which it dips.

From this it is apparent that the reason why the higher vessel empties itself into the lower is that quicksilver is heavier than water. The contrary would be the effect if the siphon, and also the vessels in which the siphon is plunged, were filled with oil, which is lighter than water, while the whole apparatus was immersed in the same tank of water. For then the oil from the lower vessel would rise and flow through the top of the siphon into the higher for the reasons aforesaid. The water would press the oil continuously from the lower vessel more strongly by the weight of one inch of height, and the oil of the longer leg would oppose this with the weight of one inch of greater height. As one inch of oil is lighter than one inch of water, the oil of the longer leg would be driven up more forcibly than that in the other and consequently would make its way from the lower vessel to the higher. And lastly, if the siphon were filled with a liquid of the same weight as the water in the tank, then there would be no exchange. Equilibrium would prevail. Calculating all the forces, you will find that they cancel one another.

Such are the effects that had to be clearly understood in order to grasp the reason for the rise of these liquids in siphons. After this it is perfectly easy to see why the weight of the air lifts the water in simple siphons from the higher to the lower vessel. We need not dilate upon the subject, as the case is only another one of the general rule that we have stated above.

VI. The weight of the mass of the air causes the flesh to rise in the process of cupping.

To explain how the weight of the air causes the flesh to rise in the process of cupping, we will discuss a wholly similar effect, produced by the weight of water, which will be entirely convincing.

This effect we have described in the Equilibrium of Liquids (Fig. XVII). A man presses against his thigh one end of a twenty-foot tube, and then takes up his position in a tank full of water, at the same time holding the upper end of the tube above the surface of the water. The flesh rises at the lower opening of the tube as if suction were applied to the spot. It is evident that this swelling is not due to the abhorrence of a vacuum: for the tube is open, and there would be no swelling if there were but little water in the tank. Obviously the weight of the water alone causes it: for this water presses the flesh in every part of the body save at the bottom end of a tube from which the water is excluded, and into which it drives the blood and the flesh that make up the swelling. What we say of the weight of water applies equally to any liquid whatsoever. If the man takes his position in a tank filled with oil, the same effect will be produced so long as the liquid touches him everywhere except in the one spot. But if the tube is removed, the swelling disappears, because the liquid now affects this spot as well as all the others, and will not impress it differently.

This being clearly understood, it will be seen to be a necessary consequence that when a candle is put on the flesh and a cupping glass on top, no sooner will the flame die than the flesh will rise. The air in the cupping glass, having been rarefied by the flame, condenses in the subsequent cooling when the flame goes out. As the air touches the body everywhere save under the cup, to the space within which it has no access, the flesh must swell there and the weight of the air must drive the blood and the adjoining pressed flesh into the unpressed spot. This occurs for the same reason, and with the same necessity as in the instance of the weight of water we gave above, when the water touched the body everywhere save in one spot. Whence it is apparent that the effect of cupping is but a particular case of the general rule concerning the action of all fluids acting upon a body which they touch everywhere save in one spot only.

VII. The weight of the mass of the air causes the attraction produced by suction.

No more than a word will now be necessary to explain why, when a man places his mouth on water and sucks, the water rises in the mouth; for we know that the weight of the air presses the water on all sides except where the mouth is set, since that is the only spot which the air does not touch. Hence when the breathing muscles expand the chest and increase the capacity within, the air inside has more room to fill than before, and less force with which to oppose the entrance of water into the mouth than the force which the outside air, pressing on that water from all sides save one, has to drive it in. That is the reason for this attraction, which is exactly the same as that of the syringes.

VIII. The weight of the mass of the air causes the flow of milk to nursing infants.

For the same reason, when an infant closes his lips round his nurse's nipple and sucks, the milk flows. The breast is pressed on all sides by the surrounding air except in the spot which is enclosed by the infant's lips. That is why, as soon as the breathing muscles increase the capacity of the infant's body, as we have said just above, and only the inside air is in contact with the nipple, the outside air, which has more force and compresses the breast, drives the milk through this opening, at which there is less opposition. This is as necessary and natural a consequence as when milk is squeezed out of the breast by the pressure of both hands.

IX. The weight of the mass of the air causes the in-drawing of air which occurs in breathing.

For the same reason, in breathing, the air enters the lungs. When the lungs open while the nose and all passages are also free and open, the air in these passages, driven by the weight of its whole mass, enters and drops down by the natural and necessary action of its weight.[6] These facts are so intelligible, easy, and simple, that it is strange that recourse should have been had to the abhorrence of a vacuum, occult qualities and other such far-fetched and chimerical causes for the purpose of accounting for them. It is just as natural for air to enter

[6] *Y entre & y tombe par l'action naturelle & necessaire de son poids.*

and drop down into the lungs when they open, as for wine to drop into a bottle when it is poured in.

Thus it is that the weight of the air produces all the effects hitherto ascribed to the abhorrence of a vacuum. I have now explained the more important of them. If any remain unexplained, they are so easy to account for by the same processes as the rest, that to look for them and explain them in detail would seem to me idle and tedious. Indeed they have all been seen, we may say, in their common origin, in the preceding treatise, since all these effects are but particular cases of the general rule concerning the Equilibrium of Fluids.

Chapter III

The Weight of the Mass of the Air being limited, so also are the Effects it produces

SINCE the weight of the air produces all the effects hitherto ascribed to the abhorrence of a vacuum, it must follow that, this weight being limited and not infinite, its effects also must be limited. Experiment confirms this, as the following will show.

Whenever the piston of a suction pump or a syringe is drawn back, the water follows it. If the action is continued, the water will still follow, but not to any height whatsoever: for there is a certain point beyond which it will not go, namely, a height of about thirty-one feet. So long as the piston is not raised beyond that height, the water will rise with it and stay there, still in close contact with the piston. But when the piston is raised any higher, the water ceases to be pulled and remains in motionless suspension at the same unchanging height. However high the piston may rise beyond, the water leaves it unaccompanied.

The reason is that the weight of the mass of the air is approximately the same as that of a height of thirty-one feet[1] of water; so that, when it drives the water up the pump by pressing upon it without and not within, to attain a balance, it raises it just to the height at which the water pressure equals its own. When this condition obtains, equilibrium is established between the water in the syringe and the air outside, just as water and quicksilver balance when their heights are in the same ratio as their weights,[2] as is shown in our Equilibrium of Liquids. And since the water does not rise excepting for the solitary reason that the weight of the air forces it to do so, it remains stationary when it reaches the particular level beyond which the weight of the air can raise it no further; there being no other force to move it.

However large the pump, the water always reaches the same level, because liquids rise not according to their bulk, but according to their height, as we have shown in the Equilibrium of Liquids.

If quicksilver is drawn up a syringe, it will rise to a height of two feet three inches and five lines, which is precisely the height at which its pressure equals that of thirty-one feet of water, because this pressure equals that of the mass of the air. And if oil is raised in a pump it will rise to about thirty-four feet or more: for at that height it weighs as much as thirty-one feet of water and consequently as much as the air. All other liquids behave in a similar manner.

When a tube sealed at the top and open at the base is filled with water, then, whatever its height may be under thirty-one feet, all the water will remain suspended in it, because the weight of the air is sufficient to hold it at such levels. But if the height exceeds thirty-one feet, enough water will fall to bring the level to thirty-one feet, where it will remain without further drop. In the same way we saw the quicksilver in a tube placed in a tank full of water drop down to the height at which its pressure equaled that of the water. But if this

[1] *Pieds,* each equal to about 1.07 English feet.

[2] *Poids* = densities. This ambiguity was not then general in scientific diction. Stevin carefully defined density. See Appendix I, Definitions II-V, p. 136.

tube were filled with quicksilver instead of water, the quicksilver would fall until it rested at the height of two feet, three inches, and five lines, which exactly corresponds to thirty-one feet of water.

And should these tubes, in which water and quicksilver have remained suspended, be slightly tilted, these liquids, being lowered by the tilting, would rise again until they reached the same height they had before. The weight of the air prevails so long as they are below that height, and is in equilibrium as soon as it is reached again. This behavior is wholly similar to that recorded in the Treatise on the Equilibrium of Liquids, in the case of a tube of quicksilver tilted in a tank full of water: when this tube is again placed erect, the liquids always rise again to the original level.

Thus it is that in a siphon all the water of the higher vessel rises and passes over into the lower so long as the branch of the siphon that dips into it has any height under thirty-one feet, because, as we have stated elsewhere, the weight of the air is fully equal to holding the water suspended at that height; but as soon as the branch in the higher vessel exceeds that height, the siphon ceases to work; that is to say, the water of the higher vessel will no longer rise to the top of the siphon and flow into the other, because the weight of the air cannot raise the water above thirty-one feet. The water at the top of the siphon, consequently, then divides, and flows down each leg into its proper vessel, but stops at the height of thirty-one feet above each vessel and stays motionless at that height, because of the countervailing weight of the air.

If the siphon is tilted somewhat, the water will rise in each leg until it reaches the level at which it stood before the tilting; and if the tilting is such that the top of the siphon is no higher than thirty-one feet above the higher vessel, the water in the leg that dips into this vessel, being at the top of the siphon, will fall down the other leg: and, if constantly replenished from the higher vessel, will flow as a thin trickle only; whereas if the tilting is increased, it will flow copiously, filling the tube.

The same effects will be produced by all other liquids, in due proportion to their weight. Thus, also, a bellows cannot be opened unless a certain degree of force is applied; but if that degree is exceeded it can be opened. Now the force required is as follows. If the sides are one foot in diameter, a force will be needed capable of raising a vessel full of water one foot in diameter like the wings of the bellows and thirty-one feet long—which is the height to which water rises in a pump. If the wings are only six inches in diameter, the force required to open the bellows will be equal to the weight of water in a vessel six inches in diameter and thirty-one feet in height, and so forth. A weight equal to that of such water, hanging from one of the wings, will open it; but a lesser weight will not, because the weight of the air that presses it is precisely equal to that of thirty-one feet of water. The same weight will draw out the piston of a sealed syringe, and will also pull apart two polished bodies laid on one another. If they are an inch in diameter, then, when a force is applied to them which is equal to the weight of a mass of water one inch in diameter and thirty-one feet in height, they will come apart.

Chapter IV

As the Weight of the Mass of the Air increases when it is more highly charged with Vapors, and diminishes when it is less so charged,[1] *so the Effects produced by its Weight increase and decrease proportionally*

SINCE the weight of the air causes the effects we are here discussing, and since that weight is not always the same in the same region but varies continually with the vapors that

[1] Although the density of moist air is less than that of dry air under the same conditions, barometric pressures at the same altitude and like temperatures are usually greater when the air is moist; primarily on account of its greater com-

blow into it, the effects cannot be at all times uniform, but must, on the contrary, be continually variable. Experiment confirms this, and shows that the thirty-one foot height which we have taken as our standard is not an invariably accurate measure. Water does not always rise in the pumps and remain in suspension at that precise height. It sometimes rises to thirty-one and one-half feet, then drops back to thirty-one; again it drops three inches down and suddenly rises one foot, according to the changes in the air. This occurs with the same irregularity with which the atmosphere clouds over or clears up.

Experiment shows that the same pump raises water higher in some weathers than in others by as much as one foot and eight inches. A pump can be built, and by the same token, a siphon also, of such a height that they will work in some weathers but not in others, according as the air is more or less charged with vapor or for some other reason weighs more or less. Should quicksilver be used instead of water, the experiment would be rather curious, and quite easy, since in this case such long tubes would not be required.

From such facts the inferences must be drawn that water in tubes remains in suspension at a lower level in certain states of the atmosphere than in others, and that a bellows is easier to open in different weathers with precisely the same variability. And so with the other effects; for what is true of one

---

pressibility at ordinary temperatures in this condition, since this may occasion an increase of density that overcompensates, at the moderate altitudes where measurements are usually made, the decrease due to the mere presence of water vapor—especially when, as is usual, the upper air is dry. Perier observed this effect in a long series of observations made with the mercury barometer at one station in 1649-51 (pp. 113 ff.); and Pascal, in his experiment with the bellows (pp. 80 ff.) not only noted it, but made it the basis of several acute meteorological inferences (p. 85. Section III: 5, 6). In 1753 Bouguer remarked especially the strikingly greater difference of pressure between two fixed altitudes when the air is moist—an informing fact which had previously been missed by Perier for lack of a sufficient number of observations; but it was not until 1784 that Deluc, as the result of a long series of measurements made under widely variable conditions and carefully compared, finally explained the phenomenon satisfactorily. See Deluc: *Recherches sur les modifications de l'atmosphere,* nouvelle ed. 1784, III, §§715, 716.

applies exactly to all the others, each according to its special nature.

Chapter V

The Weight of the Mass of the Air being greater in low places than in high, the Effects produced in the Lowlands are proportionally greater

SINCE the weight of the mass of the air produces all the effects we are examining, it ought to happen that since it is not the same at all places on the earth, inasmuch as it is greatest on the lowest ground, these effects ought also to differ. Experiment confirms this, and reveals the fact that this height of thirty-one feet which we have taken to serve as an example is not that to which water rises in pumps all over the world. The heights differ in all places which are not at the same altitude, being greater when the altitude is less, and less when it is greater. Experiments conducted at altitudes which differ by five or six hundred fathoms[1] have shown a difference of water-level of four feet and three inches. Similarly, the same pump which will raise water to a height of thirty feet and four inches at one place, at another, some five hundred fathoms higher, will not raise it under the same atmospheric conditions higher than twenty-six feet and one inch: the difference in this case being a sixth [of the lower pressure].

The same thing necessarily occurs with respect to all the other effects, each according to its character. For instance, two polished bodies are harder to separate in a valley than on a mountain top.

As a difference in altitude of five hundred fathoms causes

[1] 500 *toises* = 978 meters. Pascal's value for the corresponding difference in atmospheric pressure, given here, = 138 cm. of mercury.

a difference of four feet and three inches in the water-level, so do smaller differences in altitude cause proportionally smaller differences in water levels: one hundred fathoms, a difference of about ten inches; twenty fathoms about two inches; and so forth.[2]

The best apparatus for studying these variations is a glass tube some three or four feet long, sealed at the top and bent backward at the bottom, along the length of which a strip of paper ruled off into inches and lines is glued. If this tube is filled with quicksilver this [when the tube is placed erect] will be seen to drop some distance down, and then to remain suspended at a level that can be observed with precision. The variations of level that will follow upon changes in the air due to changes in the weather can be readily noted, as well as those which occur when the instrument is carried to a higher altitude. When it is left in one spot, the level will be seen to rise or fall as the weather changes and to be higher at one time than at another by a difference of one inch and six lines, which corresponds exactly to the one foot and eight inches of water which we have stated in the preceding chapter to be the difference due to weather. When it is carried from the foot to the top of a mountain, at a height of ten fathoms it will show a drop of close upon one line[3]; at a height of twenty fathoms a drop of two lines; of one hundred fathoms, a drop of nine lines and of five hundred fathoms a drop of three inches and ten lines. As it is carried down again, the rise of the level will go through the same gradations. All this has been verified by the experiment on the mountain Puy de Dôme in Auvergne, as will be seen in the account of that experiment which follows this treatise.[4] And these measurements of the quicksilver correspond exactly with those we have just given of water. A

[2] Pascal here ignores the effect of the self-compression of the air, of which he was aware. Cf. pp. 29 f.

[3] Corresponding to a difference of about 3 mm. in 2 meters. The following data show that the divergence of the pressure from proportionality to altitude was within the limit of errors of observation.

[4] Pages 97 ff.

corresponding explanation must be given of the resistance to the opening of a bellows and of the rest.

Whence it is apparent that precisely the same effects are produced by the weight of the air as by the weight of the water. For we have seen that when a bellows is immersed in water and is opened with difficulty because of the weight of the water, it becomes easier to open, the nearer it is brought to the surface. Also we have seen that quicksilver in a tube immersed in water remains suspended at a greater or lesser height according as it is more or less deeply immersed in the water. All these effects, whether of the weight of air or of the weight of water, follow so necessarily from the equilibrium of fluids that there is nothing more evident in the world.

Chapter VI

As the Effects of the Weight of the Mass of the Air increase or diminish with the increase or diminution of that Weight, they would cease altogether if one were above the Air or in a place where there is no Air

WE HAVE seen so far that the effects ascribed to the abhorrence of a vacuum are due in fact to the weight of the air and are always proportional to it: as that weight increases they increase, as it diminishes they diminish. For that reason, the quicksilver filling a tube remains suspended at a lower level when it is carried to a place of higher altitude, there being less air remaining above it; just as the quicksilver in a tube immersed in water falls when the tube is raised towards the surface, there being less water remaining above to balance it. Hence one may conclude with assurance that if such a tube were raised to the extreme top of the air and lifted completely out of its sphere, the quicksilver in the tube would drop entirely, there being no air at all to counterpoise it; just as the quicksilver in a tube immersed in water drops entirely when lifted bodily out of the water.

The same thing would happen if the air in the room where the experiment was conducted could be entirely removed. There being no air pressing the recurved end of the tube,[1] one must believe that the quicksilver, having no counterpoise, would fall.

But as both of these experiments are impossible, since we can neither soar above the air nor live in a room from which all the air is exhausted, it will suffice to exclude the air, not from the whole room, but only from the vicinity of the recurved end so as to keep the air from that, in order to see whether all the quicksilver will drop when the counterpoising air cannot reach it. This is easily done in the following way.

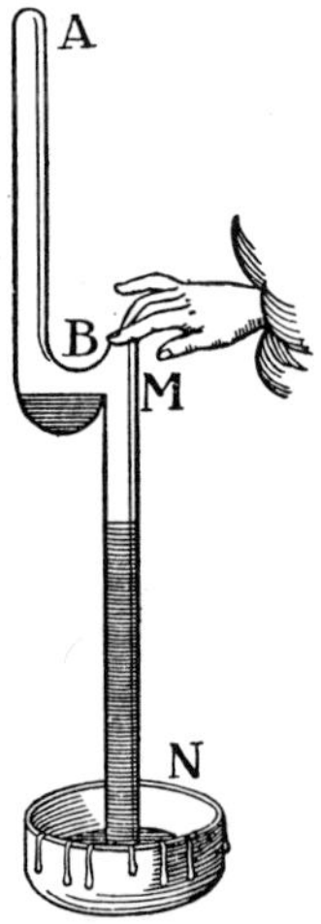

One must have a tube bent at its base, sealed at the top A and open at the bent base B, and a second tube, quite straight, open at both ends M and N, but inserted and sealed near the end M into the bend of the other as is shown in the accompanying figure.[2] The open bent end B of the first tube must be closed with a finger or otherwise, say with a pig's bladder, and the whole device turned upside down, the two tubes being now joined into one with free communication between them. If then, the whole tube having been filled with quicksilver and turned back again so as to bring the end A to the top, the end N is plunged into a bath of quicksilver, all the quicksilver from the upper tube will fall completely, and will be caught in the bend, save for a portion that will escape by the lower tube through the hole M. But the quicksilver in the lower tube will fall only in part, the rest remaining sus-

[1] Pascal here (as above) has in mind the siphon barometer.

[2] Cf. the remark (p. xxii) on an error in the drawing of this figure.

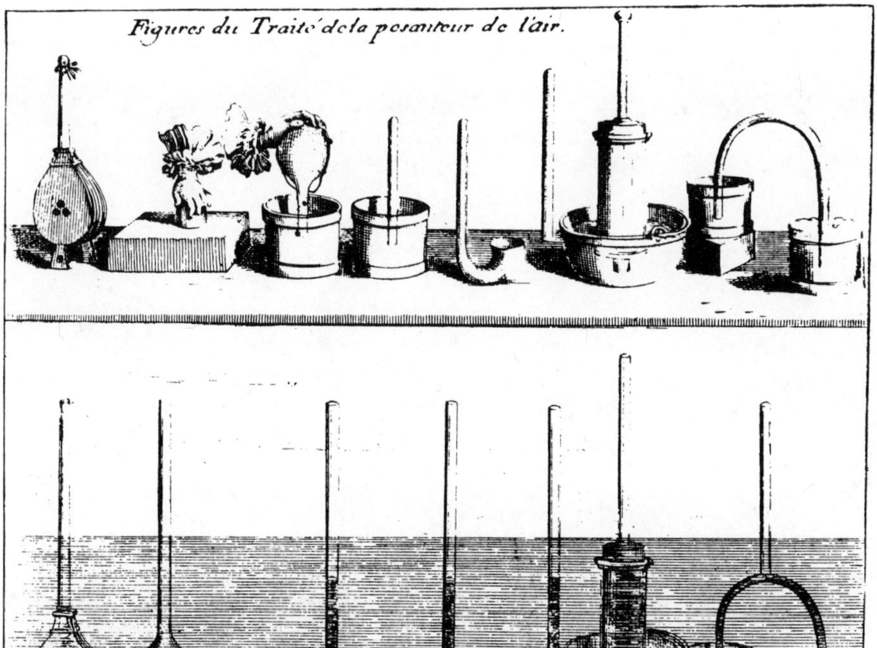
Figures du Traité de la pesanteur de l'air.

pended at a height of some twenty-six or twenty-seven inches, according to the locality of the experiment and the state of the weather.

The reason for this difference is that the air weighs on the quicksilver in the bath at the base of the lower tube and thus keeps the contained quicksilver suspended and in equilibrium; but there is no weight of air on the quicksilver at the bent end of the upper tube, since the finger or the bladder excludes the air. As there is no weight of air at that point, the quicksilver in the tube drops freely; for nothing supports it nor opposes its fall.

But nothing is ever wasted in nature: though the quicksilver in the bend does not feel the weight of the air, excluded as it is by the finger, the finger itself by compensation suffers sharply; for it bears the whole weight of the air pressing it from above, while it is unsupported from below and consequently feels as if it were pressed against the glass and drawn or sucked into the tube; and a swelling rises as if the place had been cupped. The weight of the air presses the finger, the hand, the whole body of the man on every part excepting the one spot which is at this opening, and to which it has no access; hence that spot swells and aches for the reason aforesaid.

If now the finger is lifted from this opening, the quicksilver in the bend will jump up the tube to a height of twenty-six or twenty-seven inches, because the air, falling suddenly upon the quicksilver, will drive it up instantly to the height sufficient to effect a balance. Indeed the violence of the impact will drive it even a little higher; but it will soon fall somewhat lower and then rise again until after swaying up and down a little, like a weight swinging on a thread, it will remain steady at a certain height, which is precisely that at which it balances the air.

This shows that when the air does not weigh upon the quicksilver in the bend, the quicksilver in the tube drops entirely; and consequently that if this tube had been conveyed to a place where there was no air at all, or, if such a thing were possible, above the sphere of air, the quicksilver would drop entirely.

CONCLUSION OF THE LAST THREE CHAPTERS

The conclusion is that as the weight of air is great, small, or null, so the height to which water rises in a pump is great, small, or null; and this relation is exactly proportional, like that between cause and effect. The same principle applies to the resistance encountered in opening a sealed bellows, and to the other related effects previously cited.

Chapter VII

How far Water rises in Pumps at Each Place in the World

FROM all the knowledge that we have, it follows that there are as many various levels to which the water rises in pumps as there are different places of experiment and different weathers. Thus, to the general question 'How high do suction pumps raise water?' no precise answer can be given—nor to the question 'How high do pumps raise water in Paris?', unless the state of the atmosphere be given; since the pumps raise it higher when the air is more heavily charged.[1] But we can tell how high pumps raise water in Paris when the atmosphere is most heavily charged: for in this case the conditions are all specified. Without considering the different levels to which water may be raised in each place according to the varying states of the atmosphere, however, we will take as the normal height for each place that at which the water stands when the air has a medium, or average, charge; because the measurement of this height, which stands midway between the two extremes, will yield directly a knowledge of the two others,

[1] *Chargé.* This, as the context of the following pages shows, meant laden with water vapor.

to obtain which it will be needful only to add or subtract ten inches. In this way we shall record the level to which water rises in every place in the world, however high or low it may be, when the atmosphere holds a medium charge of vapor.

But first of all it must be understood that in all pumps that are at the same altitude, water rises to exactly the same height (I mean, always, for the same state of the atmosphere); for since the air above these places has the same height, it has also the same weight and produces identical effects. For this reason we will state first the height to which water rises at places which are at sea level, because the level of the sea is everywhere the same, that is to say, equally distant from the center of the earth in every place.[2] Liquids cannot lie otherwise, or their higher parts would flow down to the lower. Thus the height to which we find water rising at any place by the seashore will be the same for all seashore places in the world, and from this it will be easy to deduce the height to which water will rise in places higher or lower by ten, or twenty, or one hundred, or five hundred fathoms, since we have determined the differences those altitudes occasion.[3]

At sea level, then, suction pumps raise the water to a height of closely thirty-one feet and two inches;[4] that is, when the atmosphere is moderately charged. Such is the measure common to all seashores in the world; whence it follows that in all such places a siphon will raise water so long as its shorter leg is of less than that height, and a sealed bellows will always open under the weight of a volume of water of that height and of a cross section[5] equal to the area of its wings.

From this point it is easy to proceed to a knowledge of the

[2] At this time it was not known that the earth is spheroidal.

[3] A first approach toward the definition of standard atmospheric pressure. The effect of changing temperature on the length of the barometric column was certainly recognized at the time, but it could not be at all closely measured, since the fixed points of the thermometric scale were not yet established. A "normal" state of the atmosphere was, of course, quite beyond any but the loosest definition, since the study of its composition was beyond possibility, excepting, perhaps, with respect to humidity.

[4] About 784 millimeters of mercury.

[5] *Largeur.*

height to which water rises in pumps in places ten fathoms higher; for since, as we have stated, ten fathoms more altitude causes a diminution of one inch in the level to which water rises, it follows that in such places water rises to thirty-one feet and one inch only.

In the same way one finds that in places twenty fathoms higher than sea level, the water rises to thirty-one feet only; in those that are one hundred fathoms above sea level the water rises to thirty feet and four inches only; in those that are two hundred fathoms above, the water rises to twenty-nine feet and six inches; in those that are some five hundred fathoms above, the water rises approximately to twenty-seven feet.

The rest can be similarly worked out. And for places below the sea level, the height to which water will rise may be ascertained by adding instead of subtracting the differences that correspond to these various heights.

CONSEQUENCES

I. From all this, it is apparent that in Paris a pump never raises water beyond thirty-two feet, nor less than twenty-nine feet and one-half.

II. It is also clear that a siphon with a short leg thirty-two feet in length will never work in Paris.

III. That a siphon with a short leg of twenty-nine feet or less, will always work in Paris.

IV. That a siphon with a short leg of exactly thirty-one feet sometimes does and sometimes does not work in Paris, according to the condition of the air.

V. That a siphon with a short leg twenty-nine feet long always works in Paris but never at a higher altitude, as for instance at Clermont in Auvergne.

VI. That a siphon ten feet tall works everywhere on the earth, since there is no mountain so high as to prevent it from working; while a siphon fifty feet in height works nowhere in the world, since there is no cave so deep that the air in it can weigh enough to raise water to such a height.

VII. That water rises in pumps at Dieppe, when the air is moderately charged, to thirty-one feet and two inches as we have said, and when the air is most heavily charged, to thirty-two feet; that it rises in pumps on mountains five hundred fathoms above sea level, when the air is moderately charged, to a height of twenty-six feet and eleven inches, and when it is least heavily charged, to twenty-six feet and one inch; that there is a difference between this height and that found at Dieppe when the air is most heavily charged, of five feet and eleven inches, which is almost a quarter of the height that is observed on mountains.

VIII. Just as we see that in all places that are at the same altitude water rises to the same height and rises less in higher places, so, conversely, if we see water rising to the same level in two different places, we may conclude that they are at the same altitude; and if water does not rise to the same level in both places, we can determine from the difference of level how much higher one is than another. This is a ready and fairly accurate way to compare the altitudes[6] of places, however far apart they may be.

Instead of a suction pump of the required height, which would be difficult to manufacture, one may use to advantage a tube four or five feet long filled with quicksilver, and sealed at the top, such as we have often described, and then observe the height at which the quicksilver remains stationary; for its height is exactly proportional to the height to which the water rises in the pumps.

IX. This also shows that degrees of heat are not correctly marked even in the best thermometers, since in them all the different heights at which the column of water stands are ascribed to the rarefaction or condensation of the air inside the tube, whereas from our experiments we learn that the

[6] *Niveler.*

changes that take place in the outer air, that is, *in the mass of the air,* contribute markedly to those changes.[7]

I pass over a number of other consequences of this new knowledge, such as, for example, the facility it affords for ascertaining the precise extent of the sphere of the air and of the vapors which are called the atmosphere:[8] for by careful observations, made at successive intervals of a hundred fathoms, of the differences registered after the first, second, and succeeding hundred fathoms, the whole height of the sphere of the air could be accurately calculated.[9]

But I leave all that in order to consider other matters more closely relevant to the subject.

[7] Pascal here refers to the air-thermoscope, the original type of dilatation thermometer, and the first instrument devised for the measurement of "degrees of heat." It was invented by Galileo in 1593 or shortly thereafter, and, according to Father Castelli, who wrote about it in 1638, consisted of a bulb of glass about the size of a hen's egg, sealed to the top of a tube the size of a straw and about two spans long, which dipped vertically into a vessel of colored water. The air in the bulb, having been warmed by the hands, later contracted as it cooled, and thus sucked colored water into the stem where its level subsequently fell and rose as the air in the bulb was warmed or cooled by changing environmental temperature. This simple device was later ingeniously modified in form but remained essentially unimproved until its most striking defect, which Pascal here points out, was recognized. It was then abandoned in favor of instruments in which the thermometric substance was liquid. The first of these was made by the French physician Jean Rey in 1638: a bulb and tube not dissimilar to those of our own common mercury thermometers but larger, unsealed, and filled with water. This instrument may have been known to Pascal; but, even if it were, it would quite naturally have been thought inferior to the air-thermometer on account of its relative insensitivity and its undependability, which was primarily due to the evaporation of its contained water. The first hermetically sealed thermometers, which contained alcohol, were made by the Florentine Academy, and were first described by them in their *Saggi*, published in 1667, five years after Pascal's death. The air-thermometer, thereafter, was not used in careful scientific work until 1703, when Amontons constructed the prototype of the instruments now employed as standards: a large bulb of air enclosed by a column of mercury by the adjustment of which it was held at constant volume, temperatures being measured by the pressures necessary to maintain this constancy, a part of which was atmospheric pressure separately determined by the barometer. (Cf. p. 109.) The phrase "degrees of heat," meaning temperature, was universally used for more than a century after this time.

[8] *L'athmosphere:* an interesting survival of the original literal meaning of this word (ἀτμός: vapor). By contrast, Pascal, when referring to the entire mass of the air, always says *la masse de l'air.* The usage indicated by this sentence, therefore, seems to have been quite definite and unequivocal.

[9] See notes on pp. 53 and 54.

Chapter VIII

How much Each Place in the World is pressed by the Weight of the Mass of the Air

FROM these experiments we learn that, since the weight of the air and that of the water in pumps balance one another, they weigh precisely the same. Therefore if we know the height to which water rises in every place in the world, we know also how each place is pressed by the weight of the air above it, and consequently that:

Places on the seashore are pressed by the weight of the air above them up to the very top of its sphere exactly as if that air were replaced by a column of water thirty-one feet and two inches high; places ten fathoms higher are pressed as if by thirty-one feet and one inch of water; places five hundred fathoms above sea-level, as if pressed by twenty-six feet and eleven inches of water; and so forth.

Chapter IX

How much the Entire Mass of All the Air in the World weighs

FROM these experiments we learn that the air at sea level weighs as much as water thirty-one feet and two inches high. But inasmuch as the air weighs less on places of higher altitude than the sea level, and therefore does not weigh the same at all places of the earth, indeed never weighs the same everywhere, it is impossible to find a fixed level which shows how much the whole earth is pressed by the air, the strong pressures compensating for the feeble.[1] But it is quite possible by conjecture to determine this level with fair approximation; as when, for instance, one is able to infer that the whole sur-

[1] *Le fort portant le foible.*

face of the earth may be thought of as pressed uniformly by the air—greater pressures compensating for lesser—when it supports a mass of water thirty-one feet high. There is certainly no error greater than half a foot of water involved in this assumption.

Now we have seen that the air over mountains that rise five hundred fathoms above sea level weighs the same as water twenty-six feet and eleven inches high. Consequently, all the air that reaches from sea level to the tops of the five-hundred fathom high mountains, weighs as much as four feet and one inch of water: which is approximately one seventh of the total mass of the air.[2]

From these same experiments we learn also that the densest vapors of the air weigh as much as a height of one foot and eight inches of water, since to balance them it is necessary to raise the water in the pumps this much above the height at which it stood when it balanced the weight of the air.[3] Thus if over any region all the vapors held by the air were condensed to water, as is the case when they turn to rain, they could not cause a rise of more than that one foot and eight inches of water in that region. And if at times storms occurred during which the rainfall reached a level higher than this, that would be because the wind carried in the vapors from neighboring regions.

We also learn from this that if the whole sphere of the air were acted upon by a force from above which compressed it against the earth into the smallest space it could possibly occupy, as if it were changed to water, it could then be no higher than just thirty-one feet. Consequently the whole mass of the air, free as it is, may be treated as though it had formerly been in a condition like that of a mass of water thirty-one feet high all round the earth, which was then rarefied and very greatly expanded until it was changed into what we

[2] See notes on pp. 53 and 54.

[3] That is, dry air; see pp. 58 f.

call air, occupying, it is true, a greater space, but conserving exactly the same weight as water thirty-one feet high.

Now nothing is easier than to compute the weight in pounds[4] of an envelope of water thirty-one feet high, which surrounds the whole earth—for a child who knew how to add and subtract could do it; and in this very way the weight in pounds of all the air in nature may be found, for it is the same. If this is done, it will be found that the air weighs close upon eight millions of millions of millions of pounds.

For the pleasure of it, I myself made the computation in this manner:

I assumed that the ratio of the diameter of a circle to its circumference is as 7 to 22, and that the diameter of a sphere multiplied by the circumference of its great circle gives its spherical surface.[5] We know that the circumference of the earth has been divided into three hundred and sixty degrees—an arbitrary division, since the circle of the earth, and that of the heavens also, could quite as well have been divided into a larger or smaller number of units,—and it has been found that each of these degrees covers 50,000 fathoms.[6]

The league in the vicinity of Paris is one of 2,500 fathoms. Hence there are 20 such leagues to a degree. Others estimate 25 leagues to a degree, but since they count but 2,000 fathoms to the league, the result is the same. There are 6 feet to a fathom. A cubic foot of water weighs 72 pounds.[7]

This being known, it is very easy to make the desired com-

[4] *Livres:* see note 7.

[5] $2R \times 2\pi R = 4\pi R^2$. (Four great circles: the original and usual Archimedean formula.)

[6] 50,000 *toises*. This distance was nearly 97,500 meters or 605.6 English miles, which corresponds to a terrestrial circumference of 21,802 miles. The estimate is therefore very rough, corresponding to some measurement much inferior to that of Eratosthenes (c.230 B.C.) though better than the guess made by Poseidonius which had been accepted by Ptolemy (150 A.D.). Actually, in Pascal's time the circumference of the earth was known to a tenth of a per cent, thanks to the exceptionally careful measurement made by the Dutch physicist Snell in 1617.

[7] *Livres.* Since a cubic foot of water weighs about 28,258 grams, this indicates that the livre in Pascal's time was about 392.5 grams, or 0.865 of the English pound.

putation. For since the earth has in its great circle or circumference . . . . 360 degrees
it consequently has a circumference of. . . . . . 7,200 leagues
and by the proportion between circumference and diameter, its diameter will be. . . . . 2,291 leagues.
Therefore, multiplying the diameter of the earth by the circumference of its great circle, we find that its total spherical surface is . . . . 16,495,200 square leagues
or . . . . 103,095,000,000,000 square fathoms
or . . . . 3,711,420,000,000,000 square feet.
And as a cubic foot of water weighs 72 pounds, it follows that a prism of water with a base of one square foot and a height of thirty-one feet weighs 2,232 pounds.

Therefore, if the earth were covered with thirty-one feet of water there would be as many thirty-one foot prisms of water as there are square feet in its whole surface. (I am well aware that these would not be prisms, but sectors of the sphere; but I purposely neglect this refinement of precision.)

Consequently the earth would bear as many times 2,232 pounds of water as there are square feet in its whole surface. Therefore that whole mass of water would be

8,283,889,440,000,000,000 pounds.

Therefore the whole mass of the sphere of air in existence is that same weight of

8,283,889,440,000,000,000 pounds,

that is to say, eight million million millions, two hundred and eighty-three thousand, eight hundred and eighty-nine million millions, four hundred and forty thousand million pounds.

CONCLUSION OF THE TWO PRECEDING TREATISES

I HAVE recorded in the preceding Treatise all the general effects which have been heretofore ascribed to nature's effort to avoid a vacuum, and have shown that it is utterly wrong to attribute them to that imaginary cause. I have demonstrated, on the contrary, by absolutely convincing arguments and experiments that the weight of the mass of the air is their real and only cause. Consequently, it is now certain that nature nowhere produces any effects in order to avoid a vacuum.

It is not difficult to demonstrate, furthermore, that nature does not abhor a vacuum at all. This manner of speaking is improper, since created nature, which is the nature under consideration, is not animated, and can have no passions. Such language is in fact metaphorical, and means nothing more than that nature makes the same efforts to avoid a vacuum as if she abhorred it. Those who use this phrase mean that it is the same thing to say that nature abhors a vacuum as to say that nature makes great efforts to prevent a vacuum. Now, since I have shown that nature does nothing at all to avoid a vacuum, the conclusion is that nature does not abhor it. To carry out the metaphor: just as we say of a man that a thing is indifferent to him when his actions never betray any movement of desire for, or of aversion to, that thing, so should we say of nature that it is supremely indifferent to a vacuum, since it never does anything either to seek or to avoid it. (I am here still using the word "vacuum" to mean a space empty of all bodies which our senses can apprehend.)

It is perfectly true (and this is what misled the ancients) that water rises in a pump when the air has no access to it, that a vacuum would result if the water did not follow the piston, and that the water ceases to rise as soon as any cracks develop by which the air can get in to fill the pump. Thus it looks as though the water rose merely for the purpose of preventing a vacuum, since it rises only when otherwise there would be one.

Similarly it is a fact that a bellows is hard to open when its apertures are so carefully sealed that no air can enter it; and it is true that its opening would produce a vacuum. This resistance ceases when air can enter to fill the bellows, and since it is met with only when a vacuum would otherwise result, it seems to be due to nothing else than the fear of a vacuum.

Finally, it is a fact that all bodies in general make great efforts to follow one another and to keep together whenever their separation, and nothing else, would produce a vacuum between them. This is why it has been inferred that this close adhesion is due to the fear of a vacuum.

To reveal the weakness of this reasoning the following example will serve. When a bellows is placed in water in the manner we have often described, with its nozzle at the end of a tube assumed to be twenty feet long which projects out of the water into the air, and with all its side apertures sealed so as to exclude the air, everyone knows that it is hard to open, and the more so the greater the amount of water above it; whereas if the vents in one of the wings are unsealed so that they admit the water freely, the resistance disappears.

If one wished to reason out this effect like the others, he might say: When the side vents are closed and when, therefore, if the bellows is to be opened, the air must enter through the tube, there is difficulty in opening it; but when water instead of air can enter to fill it, the resistance ceases. Therefore, since there is resistance only to the entrance of air, the resistance arises from an abhorrence of the air.

There is no one who will not laugh at this inference, seeing there may well be another cause of the resistance. It is evident indeed that the bellows cannot be opened without raising the water, because the water that would be pushed aside in the act of opening cannot enter the body of the bellows, and, compelled to find room for itself elsewhere, raises the whole body of the water and causes the resistance. This does not occur when the bellows has vents through which the water can enter; for whether it is opened or shut the water neither rises nor falls in consequence, since water enters the bellows just as fast as it is pushed aside, and thus offers no resistance to its opening. All this is clear, and consequently we must believe that the bellows cannot be opened without two results: first, the air does really enter, or second, the level of the water is raised. It is the latter action that causes the resistance and with this the former has nothing to do, although it occurs simultaneously.

Let us give the same explanation for the difficulty experienced in opening in the air a bellows sealed on all sides. If it were forced open two things would occur: first, a vacuum would really be formed; second, the whole mass of air would be raised and upheld. It is the latter action that causes the resistance felt; the former has nothing to do with it. This resistance also increases or diminishes in proportion to the weight of the air, as I have shown.

The same facts explain the resistance to separation offered by all bodies between which there is a vacuum: air cannot filter in, otherwise there would be no vacuum, and this being so, they cannot be separated except by raising and upholding the whole mass of the air. It is this which occasions the resistance.

Such then is the real cause of the adhesion of bodies between which there exists a possible vacuum. It was for a long time not understood because erroneous opinions were entertained which were discarded only by degrees. There have been three different periods during which different opinions of this character were held; and these involved three generally prevailing errors which made it absolutely impossible to understand the cause of this adhesion of bodies.

The first is that, in nearly all times, it was believed that the air has no weight; for the ancients said so,[1] and their professed disciples followed them blindly. They would have remained forever wedded to that theory had not keener thinkers[2] rescued them by the force of experimental evidence; for it was impossible to believe that the weight of the air causes such adherence, so long as it was held that air had no weight.

The second error lay in the belief that the elements have no weight in themselves, for the sole reason that the weight of water is not felt by those who are in it, that a bucket full of water immersed in the water is easy to lift so long as it stays there, and that its weight begins to be felt only when it is lifted out. As if these effects could not be due to another cause—or rather, as if this one was not wholly beyond all probability! For there is no sense in believing that water in a bucket has weight when out of the water, but has no weight left after it is poured back into the well; that it loses its weight when mixing with the rest, and recovers its weight when lifted above the surface. Strange are the means men employ in order to cloak their ignorance! Because they could not understand why the weight of water is not felt, and were loath to confess their ignorance, they declared it had no weight, for the satisfaction of their vanity and to the ruin of truth. Their views prevailed; and, of course, the weight of the air could not be accepted as the cause of these effects so long as this vain imagining had currency. Even had it been known that air has weight, the claim would still have been made that it has no weight when contained within itself, and consequently the be-

[1] Particularly Simplicius, the famous Peripatetic commentator (c.530 A.D.). This opinion was often wrongly attributed to Aristotle, and thus, as here, to "the ancients" in general.

[2] See p. 27, n. 3.

lief would have persisted that it can effect nothing by its weight. That is why I have shown in the Equilibrium of Liquids that water weighs the same within itself as outside, and I have explained there why, in spite of that weight, a bucket is not hard to raise while it is in the water and its weight is scarcely felt. And in the Treatise on the Weight of the Mass of the Air I have given the same demonstration in the case of the air, to clear up all doubts.

The third error is different in kind. It does not appear in connection with [the weight of] the air, but in connection with the effects which were ascribed to the abhorrence of a vacuum. Concerning these the most erroneous theories were entertained. For it had been imagined that a pump raises water not only to ten or twenty feet, which is true enough, but still farther—to a height of fifty, one hundred, or one thousand feet, or as high as you please, without limit. Likewise the belief was held that it is not only difficult, but actually impossible, to separate two polished bodies in close contact; that not even an angel, or any created force, could do so, with hundreds of exaggerations which I scorn to repeat. And so with the rest.

This is an error of observation so ancient that it cannot be traced back to its source. Heron himself, who is one of the oldest and best of the authors who have written on the raising of water, states as a positive and uncontrovertible fact that the water of a river may be made to pass over a mountain ridge and to flow into the valley beyond, provided this valley be somewhat lower down, by means of a siphon placed on the summit with its legs stretching along the slopes, one into the river and the other on the farther side; and he asserts that the water will rise from the river over the mountain and drop down again into the other valley, however high the ridge between may be.[3]

[3] No such explicit statement appears in the Schmidt edition of the *Pneumatica*, nor in the Latin edition of Commandinus (1575), which Pascal may have used, nor in another available sixteenth-century Italian edition. It is highly probable, therefore, that Pascal either misinterpreted Heron by misreading some passage like that in the *Dioptra* where Heron discusses the surveying of a

All writers on the subject have said the same thing; and even at the present time our fountain builders guarantee that they can make suction pumps which will raise water as much as sixty feet if it be desired.

Neither Heron, nor those other writers, nor the artisans, and still less the natural philosophers, can have carried their tests very far; for had they tried to draw water to the height of forty feet, they would have failed. They had only seen suction pumps and siphons six, ten, or twelve feet high, which worked beautifully; and in all the experiments they had occasion to make, had observed no case in which water failed to rise. They never imagined, consequently, that there was a limit beyond which water behaved otherwise. They conceived that the facts they had noticed were the results of an invariable natural necessity; and since they believed that water rose by an invincible abhorrence of a vacuum, they concluded that as it rose at first, so it would continue to rise without limit, applying their interpretation of what they did observe to what they did not observe and declaring both statements to be equally true.

So positively was this believed that philosophers have made it one of the most general principles of their science and the foundation of the treatises on the vacuum. It is and has been didactically asserted every day in all the schoolrooms in the world, ever since books were written. Everyone has firmly believed it, and it has remained uncontradicted down to our own time.

This fact perhaps may open the eyes of those who dare not doubt an opinion which has always been universally enter-

---

mountain top for the purpose of dropping shafts to a subterranean water channel (*Heronis Alexandrini opera quae supersunt omnia,* III, ed. Schöne, 1903, *Dioptra* 16, p. 243), or based his statement on hearsay. On the other hand, Heron, in discussing the siphon (*Ibid.,* I, ed. Schmidt, 1899, *Pneumatica* 2, p. 32) makes no mention of any limitation to its efficacy; neither does Cardan (*De subtilitate*, 1560: I. 3.364a); and Galileo's discussion of pumps in the *Mathematical Discourses* of 1638 (see below, Appendix II, pp. 159 ff.) implies that it was then still taken for granted by the philosophers, if not by all workmen, that a siphon of any height would operate.

tained; for simple workmen have been able to prove in this instance that all the great men we call philosophers were wrong. Galileo declares in his dialogues[4] that Italian plumbers taught him that water rises in pumps only to a certain height: whereupon he himself confirmed the statement as others did also, afterward, first in Italy and later in France, by using quicksilver, which is easier to handle but provides merely several other ways of making the same demonstration.

Before men gained that knowledge, there was no incentive to prove that the weight of the air was the cause of water rising in pumps; since, the weight of the air being limited, it could not produce an unlimited effect.

But all these experiments were insufficient to show that the air does produce those effects: they had rid us of one error but left us in another. They taught us, to be sure, that water rises only to a certain height, but they did not teach us that it rises higher in low-lying places. On the contrary, the belief was held that it always rises to the same height, in every place on the earth. And since the weight of the air never entered anybody's head, it was vaguely thought that the nature of the pump was such that it lifted water to a limited height and no further. Indeed Galileo took that to be the natural height of a pump, and called it *la altessa limitatissima*. How indeed could it have been imagined that that height was different in different places? Certainly, it would seem improbable. Yet that last error again put out of the question the proof that the weight of the air causes these effects; since, because this weight would be greater at the foot than at the top of a mountain, its effects, obviously, would be proportionately greater there.[5]

[4] See Appendix II, p. 161.

[5] Pascal here refers, in the first place, to ideas current up to the time of Galileo, which were consistent with the theory of a limitless *horror vacui*; then to Galileo's observations on the pump which, though they permitted the retention of the general notion of a *horror vacui*, nevertheless proved that it was a finite and measurable force. But Galileo did not dogmatically assert, as Pascal implies, that it was an invariable force: it was contrary to his habit of thought, in the absence of experimental evidence, to do anything of the sort; indeed, for lack of sufficient knowledge, he left the phenomenon unexplained. Pascal's omission of all reference to the several physicists who, after Galileo,

That is why I decided that the proof could be obtained only by experimenting in two places, one some four or five hundred fathoms above the other. I chose for my purpose the Puy de Dôme mountain in Auvergne, for the reasons that I have set forth in a little paper which I printed as early as the year 1648, immediately after the experiment had proved successful.[6]

This experiment revealed the fact that water rises in pumps to very different heights, according to the variation of altitudes and weathers, but is always in proportion to the weight of the air. It perfected our knowledge of these effects, and put an end to all doubting; it showed their real cause, which was not abhorrence of a vacuum, and shed on the subject all the light that could be wished for.

Try now to explain otherwise than by the weight of the air why suction pumps do not raise water so high by one-quarter on the top of Puy de Dôme in Auvergne as at Dieppe; why the same siphon lifts water and draws it over at Dieppe and not at Paris; why two polished bodies in close contact are easier to separate on a steeple than on the street level; why a completely sealed bellows is easier to open on a house-top than in the yard below;[7] why, when the air is more heavily

quite definitely conceived that this force was external and due in fact to atmospheric pressure—among whom were Mersenne and Descartes, whose ideas were almost certainly known to him—and particularly his failure to mention Torricelli, who verified this theory, leaves a false impression too favorable to himself. It was not his work that invalidated the conception of *horror vacui*: and his contradictory hypotheses were not only suggested but were almost completely verified by his predecessors. It is sufficient honor to him that he so brilliantly completed their work.

[6] This appears in the sequel (pp. 97 ff.).

[7] In this illustration the difference of level would probably not be less than 10 *toises* or 64 feet (p. 109), which, according to Pascal's measurements (pp. 59 ff.), would cause a difference of level in the water barometer of one *poulce*, or about 2.7 centimeters. Thus it would be easily perceptible with this instrument, though not, in Pascal's time, with a barometer of mercury, which would have shown a difference of about 1/15 *poulce*, or less than one *ligne*, which was close to the limit of precision in observation (p. 104). If the bellows in this illustration—or the polished bodies in the preceding—had the lateral area of a square *pied*, or roughly 1,000 square centimeters, a difference of 10 *toises* in height would change the force required to open it by the weight of 2.7 x 1,000 grams or 2.7 kilograms. Even a twentieth of this force—which would roughly

charged with vapors, the piston of a syringe is harder to withdraw;[8] and lastly why all these effects are invariably proportional to the weight of the air, as effects are to their cause.

Does nature abhor a vacuum more in the highlands than in the lowlands? In damp weather more than in fine? Is not its abhorrence the same on a steeple, in an attic, and in the yard? Let all the disciples of Aristotle collect the profoundest writings of their master and of his commentators in order to account for these things by abhorrence of a vacuum if they can. If they cannot, let them learn that experiment is the true master that one must follow in Physics; that the experiment made on mountains has overthrown the universal belief in nature's abhorrence of a vacuum, and given the world the knowledge, never more to be lost, that nature has no abhorrence of a vacuum, nor does anything to avoid it; and that the weight of the mass of the air is the true cause of all the effects hitherto ascribed to that imaginary cause.

---

measure the corresponding difference for a small bellows with circular wings three inches in diameter, such as Pascal actually used (p. 81)—would be nearly five ounces, and easily observable.

[8] See note on pp. 51 f.

FRAGMENTS OF OTHER WORKS BY PASCAL

FRAGMENT

Of another longer Work by M. Pascal on the same subject, divided into Parts, Books, Chapters, Sections, and Articles, of which only the following were found among his papers.

Part I, Book III, Chapter I, Sections II [and III]

SECOND SECTION

The Effects vary according to the variations of the Weather, and are more or less marked in proportion as the Air is more or less charged [1]

WE HAVE seen, in the Introduction on the subject of the weight of the air, that in the same region the air weighs more at one time than at another, according as it is more or less charged. And we will show in this section that the effects vary in the same region as the weather varies, and that they are more or less marked as the air is more or less charged.

ARTICLE I

To make the determination of this variation accurate, it is necessary to have a glass tube such as we have several times described, sealed at the upper end, open and recurved at the lower end, and full of quicksilver which remains suspended at a certain height. Let this tube be set up permanently in a

[1] Charged with moisture, humid—as implied below.

room where it can be readily examined and will not be tampered with. Let a strip of paper, ruled off in inches and lines, be glued along its length so that one can read the exact point at which the quicksilver stands suspended, as is done with thermometers.

At Dieppe, when the weather is most markedly overcast, the quicksilver will stand at the height of twenty-eight inches and four lines above its level in the bent end; but when it clears, it will be seen to drop, possibly by four lines. The next day, it may be seen to have dropped ten lines; sometimes an hour later it will have gone up ten lines and sometimes thereafter it will have gone up or down according as the weather is more or less overcast. Between the two extremes there will be found a difference of eighteen lines, that is to say, the quicksilver will be sometimes at a height of twenty-eight inches and four lines, and at other times at a height of twenty-six inches and ten lines.

This experiment is called the "continuous experiment," because one may observe, if he wishes, continuously; and the quicksilver will be found at almost as many different heights as there are different weather conditions at the times of observation.

ARTICLE II

The perfect conformity of all the effects ascribed to abhorrence of a vacuum being such that what is said of one applies to all the others, we may conclude with certainty that since the height of the suspended mercury varies with the weather, there will be similar variation in all the other effects: as in the height to which pumps raise water, which is likewise greater in some weathers than in others, and in the resistance of a sealed bellows to being opened, which is also greater in some weathers than in others, and so forth.

In case anyone wishes to enjoy some of these other experiments, we will indicate here, by describing that with the sealed bellows, how he may proceed.

Take a bellows, smaller than the usual sort, of which the sides are only three inches in diameter. Seal it hermetically everywhere. Fasten one of the sides to the ceiling beam in a room. To the other side attach an iron chain the links of which not only reach, but trail on, the floor. The size of the chain and the distance from ceiling to floor must be such that the links hanging from the bellows to the floor—exclusive of those that trail on the floor—weigh approximately 120 pounds. It will be found that this weight will open the bellows, since it takes but a weight of 113 pounds to do so, as we said in Book II, Chapter I, Article I.

As the bellows opens, the side to which the dragging chain is attached will come down; therefore the chain itself will drop, and the links that previously hung nearest to the floor will rest on the floor so that their weight will no longer pull upon the bellows; and the wider the bellows opens, the fewer links will remain hanging. Therefore, when the bellows is so widely opened that there are only enough links hanging to weigh 113 pounds, then if the weather is very much overcast and the air is heavily charged with vapor, the chain will drop no farther, the bellows will remain partly open, the chain partly hanging and partly trailing, and the whole device will be stationary.

And the most amazing thing is this: that when the weather clears, and when consequently a lesser weight will suffice to open the bellows, the hanging links, which weigh 113 pounds and were in equilibrium with the air when it was most heavily charged, will become too heavy because the air has cleared. They will therefore pull down the side of the bellows and open it wider until the links that remain suspended balance the weight of the air above in its new condition; and the lighter the air becomes the more links will drop. But when the air charge increases, the bellows will, on the contrary, be seen to close up seemingly of its own accord, and doing so, will pull the chain up again, until the hanging links are once more in equilibrium with the weight of the air above in this new condition. Thus the chain will rise and fall, and the bellows will open or close more or less as the air charges or discharges itself; and at all times the hanging links will be in equilibrium with the outer air. This air, surrounding and pressing the bellows on every side, would hold it shut but for the chain striving to open it; and on the other hand the chain would hold it permanently open were not the air striving to close it. But these two opposite strivings balance one another, as we have stated.

It remains to be said that when the air charge is heaviest the hanging links weigh 113 pounds, and when it is lightest, only 107 pounds; and these two limits of fluctuation,[2] 113 pounds and 107 pounds, are in perfect harmony with the two corresponding limits of the height of suspended mercury—namely, twenty-eight inches and four lines and twenty-six inches and ten lines: for a cylinder of mercury which, like the sides of the bellows, is three inches in diameter and twenty-eight inches and four lines high, weighs 113 pounds; while one of the same diameter and twenty-six inches and ten lines high weighs 107 pounds.

ARTICLE III

If you wish to get yet greater enjoyment out of these observations, you must carry out three or four such operations at once. For instance, you ought to have a tube filled with mercury such as was described in Article I, a sealed bellows, like that described in Article II, a suction pump 35 feet long, and a siphon with a short leg about 31 feet high and a long leg 35 feet high.

If you observe all these effects simultaneously, you will find that in the most heavily charged weather, the mercury in the tube will stand at 28 inches and 4 lines, and the links hanging from the bellows will weigh 113 pounds; the water level in

[2] *Ces deux mesures periodiques.*

the pump will be 32 feet, and the siphon will work, since its short leg will be under 32 feet.

And when the weather clears up somewhat, the mercury will be lowered by 12 lines and will reach only to 27 inches and 4 lines. In the same proportion, the chain will have no hanging links beyond a weight of 109 pounds; the water in the pump will be lowered a foot and therefore will be only 31 feet high; the siphon will deliver only a trickle of water, since the short leg will be exactly 31 feet high.[3]

And when the weather is clearest, the mercury will be lowered 18 lines and will reach a height of only 26 inches and 10 lines; the hanging links will weigh only 107 pounds; the water will be lowered one foot and six inches and will reach a height of only 30 feet and 4 inches. The siphon will stop working because its [vertical] shorter leg, which is 31 feet long, exceeds 30 feet and 4 inches, at which height water in that weather stops rising in a pump; but the water will hang in both legs of the siphon at that same height of 30 feet and 4 inches, as in the pump, according to the rule of the siphon.

Sometime later the mercury, the chain, and the water will rise again, and a trickle of water will flow in the siphon. Later again all levels will be depressed, then will rise again, and each time all the devices will show the same differences, and the performance will continue as long as you wish to enjoy it.

If the water siphon is in a yard on the ground level and the mercury in one of the rooms of a house, whenever you see the mercury rising in the room, you may be sure, without seeing it, that the siphon is working in the yard outside. And when you see the quicksilver drop, you may be sure, without seeing it, that the siphon has stopped, because all the effects work together, in instant response to the weight of the air which governs them all, each according to its own special character.

[3] At this point the author has a marginal note, as follows: "These experiments with the pump and the siphon may be made with greater ease if one uses mercury instead of water, as indicated in the Treatise on the Weight of the Mass of the Air."

Third Section

On the rule for the Variations in these Effects, due to Variations in the Weather

SINCE the variations in these effects are due to variations in the condition of the air, and since the latter are very extraordinary[4] and almost without any regularity, the variations in the effects are also so strange that it is difficult to bring them under any rule. We shall note, however, all that we have found of greater certainty or regularity among these effects, in the attempt to explain them all, in the usual way, by one alone, such as the suspension of mercury in a tube sealed at the top, an effect of which we have made frequent use.

1. There is a certain upper limit and also a certain lower limit which the height of the mercury almost never exceeds, because there is a certain range [of concentration] in the charge of the air which is almost never exceeded, either in weather when the atmosphere is as clear as it ever is seen to be, or when it is as overcharged as it can possibly be. Not but that there may occur some accident in the air, which would charge it excessively, in which case the mercury would rise exceptionally high; but such an occurrence is so rare that one cannot explain it by rule.

2. The mercury is seldom seen at either of its limiting heights, and ordinarily it is between the two: sometimes nearer to the one, and sometimes nearer to the other, because it happens rarely that the atmosphere is either wholly clear or charged to capacity; usually it is only partly charged, sometimes more and sometimes less.

3. These changes are irregular both in the height of the mercury and in the condition of the air; so that in a quarter of an hour, sometimes, a large difference is noticeable, while

[4] *Bizarres.*

sometimes the difference is very slight for a space of four or five days.

4. The season when the mercury is usually highest is winter; it is ordinarily lowest in summer. It is least variable at the solstices, and most variable at the equinoxes. Not but that the mercury is sometimes high in summer, low in winter, irregular at the solstices, and steady at the equinoxes, for there is no regularity in the matter; but usually things go as we have said, because, likewise usually, though not always, the air is most heavily charged in winter, and least in summer, most variably in March and September, and most constantly at the solstices.[5]

5. It is also usual for the mercury to fall in fine weather, and to rise when the weather is cold or overcast; but such is not invariably the case, for it rises sometimes when the weather is clearing, and falls occasionally when the skies cloud over. The reason is, as we have stated in the Introduction, that sometimes when the weather improves in the low regions, the atmosphere, nevertheless, considered with reference to all regions together, increases in weight; or, stated otherwise, although the air becomes charged at lower altitudes, it sometimes loses its charges elsewhere.

6. But it is another very remarkable fact that when the air becomes cloudy and the mercury falls at the same time, one may be sure that the clouds at lower altitudes are tenuous and will soon be dispersed, and that fine weather is soon coming. On the other hand, when the weather is fine and the mercury at the same time is high, one may be equally confident that there are scattered vapors about, though unseen, which will condense before long into rain. Also when low mercury and fair weather are seen together, it may be expected that fine weather will last, because there are but light charges in the air. And finally, when heavily charged air and high mercury are seen together, it may be predicted that bad weather will last, because undoubtedly the air is heavily charged. Not

[5] In the text *equinoxes,* an obvious slip.

but that a sudden wind may blow in and bring to naught these forecasts; but ordinarily they are fulfilled, because the height of the suspended mercury, being an effect of the existing air charge, is also an unequivocal indication of this charge, and one incomparably more trustworthy than those of the thermometer or any other instrument.

This knowledge may be very useful to farmers, travelers, and others for ascertaining the present state of the weather and its immediate promise, though not for telling the weather three weeks hence. But I shall consider no further the advantages that may be derived from this new knowledge, in order to proceed with our program.

ANOTHER FRAGMENT

On the same Subject, consisting of Tables, of which only seven were found, which bear the following Captions

Notice

FOR the proper understanding of these tables it must be borne in mind:

I. That Clermont is the capital city of Auvergne, the altitude of which is estimated to be 400 fathoms above Paris.

II. That Le Puy is a mountain in Auvergne in the immediate vicinity of Clermont, called Le Puy de Dôme, and about 500 fathoms higher than Clermont.

III. That Lafon is a place called Lafon de l'Arbre, on the flank of the mountain Le Puy de Dôme, much nearer its base than its summit, in fact, but assumed in the following tables, nevertheless, to be the mean height of the mountain, and consequently equally distant from its base and its summit, that is, approximately 250 fathoms from each.

It must be known also that *Pa.* or *Par.* stands for *Paris*, *Cler.* or *Clerm.* for *Clermont, Laf.* or *Lafo.* for *Lafon, Le Pu.* for *Le Puy, med.* for *medium, diff.* for *difference, ft.* for *feet, in.* for *inches*, *li.* for *lines*, *lbs.* for *pounds*, and *oz.* for *ounces.*

Second Table

To determine the weight of a leaden cylinder, which shall be equal to the resistance offered by two polished surfaces in contact, when the attempt is made to separate them.

The resistance in question is equal to the weight of a leaden cylinder, having for its base the common surface of the bodies, and for its height, *when the atmosphere is charged*:

	most heavily			*moderately*			*least*			*difference*	
	ft.	in.	li.	ft.	in.	li.	ft.	in.	li.	in.	li.
At Paris	2.	9.	4.	2.	8.	6.	2.	7.	8.	1.	8.
Clermont	2.	6.	10.	2.	6.	...	2.	5.	2.	1.	8.
Lafon	2.	5.	2.	2.	4.	4.	2.	3.	6.	1.	8.
Le Puy	2.	3.	6.	2.	2.	8.	2.	1.	10.	1.	8.

DIFFERENCES

between one place and another
when the atmosphere is charged:

	most heavily		*moderately*		*least*	
	in.	li.	in.	li.	in.	li.
Between Par. and Clerm.	2.	6.	2.	6.	2.	6.
Clerm. and Laf.	1.	8.	1.	8.	1.	8.
Laf. and Le Puy	1.	8.	1.	8.	1.	8.
Cler. and Le Puy	3.	4.	3.	4.	3.	4.
Paris and Le Puy	5.	10.	5.	10.	5.	10.

Third Table

To determine the force necessary to separate two bodies in contact on a surface with a diameter of one foot *when the atmosphere is charged:*

	most heavily lbs.	*moderately* lbs.	*least* lbs.	*difference* lbs.
At Paris	1808.	1761.	1714.	94.
Clerm.	1675.	1628.	1581.	94.
Lafon	1579.	1532.	1485.	94.
Le Puy	1483.	1436.	1389.	94.

DIFFERENCES

between one place and another when the atmosphere is charged:

	most heavily lbs.	*moderately* lbs.	*least* lbs.
Between Par. and Clerm.	133.	133.	133.
Clerm. and Laf.	96.	96.	96.
Laf. and Le Puy	96.	96.	96.
Cler. and Le Puy	192.	192.	192.
Par. and Le Puy	325.	325.	325.

Fourth Table

To determine the force necessary to separate two bodies in contact on a surface with a diameter of six inches *when the atmosphere is charged:*

	most heavily		*moderately*		*least*		*difference*	
	lbs.	oz.	lbs.	oz.	lbs.	oz.	lbs.	oz.
At Paris	452.	...	440.	4.	428.	8.	23.	8.
Clerm.	419.	6.	407.	10.	395.	14.	23.	8.
Laf.	395.	10.	383.	14.	372.	2.	23.	8.
Le Puy	371.	14.	360.	2.	348.	6.	23.	8.

DIFFERENCES

between one place and another when the atmosphere is charged:

	most heavily		*moderately*		*least*	
	lbs.	oz.	lbs.	oz.	lbs.	oz.
Between Par. and Clerm.	32.	10.	32.	10.	32.	10.
Cler. and Laf.	23.	12.	23.	12.	23.	12.
Laf. and Le Puy	23.	12.	23.	12.	23.	12.
Clerm. and Le Puy	47.	8.	47.	8.	47.	8.
Par. and Le Puy	80.	2.	80.	2.	80.	2.

FIFTH TABLE

To determine the force necessary to separate two bodies in contact on a surface with a diameter of one inch *when the atmosphere is charged:*

	most heavily lbs.	oz.	*moderately* lbs.	oz.	*least* lbs.	oz.	*difference* oz.
At Paris	12.	9.	12.	4.	11.	15.	10.
Clerm.	11.	11.	11.	6.	11.	1.	10.
Laf.	11.	1.	10.	12.	10.	7.	10.
Le Puy	10.	7.	10.	2.	9.	13.	10.

DIFFERENCES

between one place and another when the atmosphere is charged:

	most heavily lbs.	oz.	*moderately* lbs.	oz.	*least* lbs.	oz.
Between Par. and Cler.	...	14.	...	14.	...	14.
Cler. and Laf.	...	10.	...	10.	...	10.
Laf. and Le Puy	...	10.	...	10.	...	10.
Clerm. and Le Puy	1.	4.	1.	4.	1.	4.
Par. and Le Puy	2.	2.	2.	2.	2.	2.

Sixth Table

To determine the force necessary to separate two bodies in contact on a surface with a diameter of six lines *when the atmosphere is charged:*

	most heavily		*moderately*		*least*		*difference*
	lbs.	oz.	lbs.	oz.	lbs.	oz.	oz.
At Paris	3.	1.	3.	...	2.	15.	2.
Clerm.	2.	12.	2.	11.	2.	10.	2.
Laf.	2.	9.	2.	8.	2.	7.	2.
Le Puy	2.	6.	2.	5.	2.	4.	2.

DIFFERENCES

between one place and another when the atmosphere is charged:

	most heavily	*moderately*	*least*
	oz.	oz.	oz.
Between Par. and Cler.	5.	5.	5.
Cler. and Laf.	3.	3.	3.
Laf. and Le Puy	3.	3.	3.
Cler. and Le Puy	6.	6.	6.
Par. and Le Puy	11.	11.	11.

Seventh Table

To determine the height to which mercury or quicksilver rises, and remains suspended, in the common experiment, when the atmosphere is charged:

	most heavily			*moderately*			*least*			*difference*	
	ft.	in.	li.	ft.	in.	li.	ft.	in.	li.	in.	li.
At Paris	2.	4.	4.	2.	3.	7.	2.	2.	10.	1.	6.
Clerm.	2.	2.	3.	2.	1.	6.	2.	...	9.	1.	6.
Laf.	2.	...	9.	2.	...	...	1.	11.	3.	1.	6.
Le Puy	1.	11.	3.	1.	10.	6.	1.	9.	9.	1.	6.

DIFFERENCES

between one place and another when the atmosphere is charged:

	most heavily		*moderately*		*least*	
	ft.	in.	ft.	in.	ft.	in.
Between Par. and Clerm.	2.	1.	2.	1.	2.	1.
Cler. and Laf.	1.	6.	1.	6.	1.	6.
Laf. and Le Puy	1.	6.	1.	6.	1.	6.
Clerm. and Le Puy	3.	...	3.	...	3.	...
Par. and Le Puy	5.	1.	5.	1.	5.	1.

Eighth Table

To determine the height to which water rises, and remains suspended, in the common experiment, when the atmosphere is charged:

	most heavily		*moderately*		*least*		*difference*	
	ft.	in.	ft.	in.	ft.	in.	ft.	in.
At Paris	32.	...	31.	2.	30.	4.	1.	8.
Clerm.	29.	8.	28.	10.	28.	...	1.	8.
Laf.	28.	...	27.	2.	26.	4.	1.	8.
Le Puy	26.	3.	25.	6.	24.	7.	1.	8.

DIFFERENCES

between one place and another when the atmosphere is charged:

	most heavily		*moderately*		*least*	
	ft.	in.	ft.	in.	ft.	in.
Between Par. and Clerm.	2.	4.	2.	4.	2.	4.
Clerm. and Laf.	1.	8.	1.	8.	1.	8.
Laf. and Le Puy	1.	8.	1.	8.	1.	8.
Clerm. and Le Puy	3.	4.	3.	4.	3.	4.
Par. and Le Puy	5.	8.	5.	8.	5.	8.

EXPERIMENTS ON THE EQUILIBRIUM OF FLUIDS

NOTICE

Among the papers of M. Pascal was also found a printed account, dated 1648, of the celebrated experiment conducted in that year on the Puy de Dôme mountain in Auvergne, which it has seemed fitting to add to the two foregoing Treatises because it is very helpful in understanding them, and also because it is no longer in print.

The Treatise mentioned several times in this narrative is a great Treatise that M. Pascal had written on the Vacuum, of which the few fragments that have been found are reproduced in the foregoing pages.

The foot used in this experiment consists of twelve standard inches.

STORY OF THE GREAT EXPERIMENT ON THE EQUILIBRIUM OF FLUIDS[1]

Devised by Monsieur B. Pascal

in pursuance of the completion of the Treatise promised in his shorter work[2] on the Vacuum

And carried out by Monsieur F. P.[3] on one of the highest mountains in Auvergne, commonly known as Le Puy de Dôme.

AT THE *time when I published my pamphlet entitled* "New Experiments touching the Vacuum, etc.," *I used the phrase 'Abhorrence of the Vacuum' because it was universally accepted and because I had not then any convincing evidences against it;[4] but nevertheless sensed certain difficulties which made me doubt the truth of that conception. To clarify these doubts, I conceived at that very time the experiment here described, which I hoped would yield definite*

[1] *Liqueurs,* as elsewhere above.
[2] *Abbregé,* see p. xvii n.
[3] Perier.
[4] *De preuves convaincantes du contraire.*

knowledge as a ground for my opinion. I have called it the Great Experiment on the Equilibrium of Fluids, because it is the most conclusive of all that can be made on this subject, inasmuch as it shows the equilibrium of air and quicksilver, which are, respectively, the lightest and the heaviest of all the fluids known in nature.

But since it was impossible to carry out this experiment here, in the City of Paris, and because there are very few places in France that are suitable for this purpose and the town of Clermont in Auvergne is one of most convenient of these, I requested my brother-in-law, M. Perier, counsellor in the Court of Aids in Auvergne, to be so kind as to conduct it there. What my difficulties were and what the experiment is will be made clear by the accompanying letter concerning it which I wrote to him at the time.

Copy of the letter of Monsieur Pascal the Younger to Monsieur Perier, November 15, 1647

MONSIEUR,

I SHOULD not break in upon the constant calls made on you by your official duties to submit to you considerations of Physical Science, were I not fully aware that they will be a refreshment to you in your hours of relaxation, and will be as entertaining to you as they would be burdensome to others. I hesitate the less to do this, also, because I know full well the delight you take in these pursuits. You will find here but a continuation of our former discussions concerning the vacuum. You know the views of the Philosophers on this subject. They have all endorsed the principle that nature abhors a vacuum, and most of them have gone further and maintained that nature cannot admit of it, and would perish sooner than suffer it. Thus opinions have been divided: some have been content to say only that nature abhors a vacuum, others have maintained that she could not tolerate it. I have tried in my pamphlet on the vacuum to refute the latter opinion, and I believe that the experiments recorded there suffice to show indubitably that nature can, and does, tolerate any amount of space empty of any of the substances that we are acquainted with, and that are perceptible to our senses. I am now engaged in testing the truth of the former statement, namely, that nature abhors a vacuum, and am trying to find experimental ways to show whether the effects ascribed to the abhorrence of a vacuum are really attributable to that abhorrence, or to the weight and pressure of the air. For, to reveal to you frankly my whole thought on the matter, I can hardly admit that nature, which is not at all animated nor sensible, can be capable of abhorrence, since the passions presuppose a soul capable of experiencing them. I feel much more inclined to attribute all these effects to the weight and pressure of the air, because I consider them only as particular cases of a universal principle concerning the Equilibrium of Fluids, which is to be the greater part of the Treatise I have promised. Not but that I had the same thoughts when I brought out my abridgment; but for lack of convincing experiments, I dared not then (and I dare not yet) give up the idea of the abhorrence of a vacuum. I even used it as a premise in my abridgment, not having then any other design than to controvert the opinion of those who hold that the void is absolutely impossible, and that nature would rather suffer her destruction than the least empty space. Indeed I do not consider that it is permissible for us lightly to discard the maxims handed down to us by the ancients unless we are compelled to do so by indubitable and unanswerable proofs. But in that case I maintain that it would be the extremity of weakness to have the least scruple in the matter. We must have more respect for evident truths than obstinacy in clinging to accepted opinions. I cannot better illustrate the circumspection which I exercise before rejecting ancient maxims than to recall to you the experiment I made lately in your presence with the two tubes, one inside the other, which exhibit a vacuum within a vacuum.[5] You saw that the quicksilver of the inner tube hung suspended at the usual height when it was counterpoised and pressed by the weight of the whole mass of the air, but that it dropped altogether, so that it was no longer suspended at all, when by removing all the surrounding air we made a complete vacuum about it so that it was no longer pressed by it and counterbalanced. Afterward, you saw that the height or suspension of the quicksilver increased or decreased as the pressure of the air increased or decreased, and finally that all these various heights or suspensions of the quicksilver were always proportional to the pressure of the air.

After that experiment we certainly had reason to believe, as we do believe, that it is not the abhorrence of the vacuum that causes the quicksilver to stand suspended in the usual experiment, but really the weight and pressure of the air, which balances the weight of the quicksilver. But seeing that all the effects of this last experiment with the two tubes, which is so naturally explained by the mere pressure and weight of the air, can also be explained, probably enough, by the abhorrence of a vacuum, I still hold to that ancient principle, although I am determined to seek a thorough elucidation of this difficulty by means of a decisive experiment.

To this end I have devised one that is in itself sufficient to give us the light we seek if it can be carried out with accuracy.

[5] This experiment was probably suggested by Torricelli's remark in his second letter to Ricci (Appendix III, p. 168) that, since a diminution of external pressure reduced the height of the barometric column, this column would be reduced to nothing if the surrounding space were vacuous. The test was carried out in 1647. A glass tube three feet long, completely filled with mercury and closed at both ends with membranes, was supported like the barometric tube with the lower end dipping into a vessel of mercury, and was then hung within and near the upper end of a larger tube which was likewise filled with mercury and closed by membranes. When this whole apparatus was set up vertically in a second dish of mercury, and the basal membrane of the outer tube was broken, this tube functioned like a barometer and the liquid in it fell far enough to leave the included tube and dish suspended *in vacuo*. When, now, the basal membrane of this smaller tube was broken, the mercury within it fell to the level in its basin; but when the *upper* membrane of the larger surrounding tube was punctured, letting air into this tube, the mercury rose to the barometric level in the inner tube.

This is to perform the usual experiment with a vacuum several times over in one day, with the same tube and with the same quicksilver, sometimes at the base and sometimes at the summit of a mountain at least five or six hundred fathoms high, in order to ascertain whether the height of the quicksilver suspended in the tube will be the same or different in the two situations. You see at once, doubtless, that such an experiment is decisive. If it happens that the height of the quicksilver is less at the top than at the base of the mountain (as I have many reasons to believe it is, although all who have studied the matter are of the opposite opinion), it follows of necessity that the weight and pressure of the air is the sole cause of this suspension of the quicksilver, and not the abhorrence of a vacuum: for it is quite certain that there is much more air that presses on the foot of the mountain than there is on its summit, and one cannot well say that nature abhors a vacuum more at the foot of the mountain than at its summit.

But since difficulty, as a rule, attends great achievement, I foresee much trouble in carrying out this plan, because for the purpose a very high mountain must be selected, in the vicinity of a town where a person may be found who is competent to bring to bear upon the task all the precision of measurement that it demands. If the mountain were very distant, it would be difficult to carry to it the vessels, the quicksilver, the tubes, and many other necessary accessories, and to undertake the many laborious journeys that would be necessary in order to find upon these heights the suitably calm weather which is seldom to be met with there. And since it is as uncommon to find outside of Paris persons who have these qualifications as it is to find places that meet the conditions, I have been highly gratified by my good fortune in having on this occasion found both; for our town of Clermont is at the foot of the lofty Puy de Dôme, and I hope that you will be good enough to grant me the favor of conducting the experiment yourself. Being assured of this, I have encouraged all our interested Parisians to expect it, among others the Revd. Father Mersenne, who has already pledged himself by letters to Italy, Poland, Sweden, Holland, and elsewhere, to convey the result to the friends his great merit has won him in those countries. I say nothing about the means of performing the experiment, because I well know that you will omit none of the precautions necessary to carry it out with precision.

I would only beg of you to choose the earliest date you can, and to excuse the liberty I am taking, to which I am driven by my impatience to hear of the success of the experiment. Without it I am unable to put the finishing touches to the Treatise I have promised the public, or to give satisfaction to the many persons who are impatiently awaiting it, and who will be infinitely grateful to you. Not that I wish to lessen my own gratitude by sharing it with so many; it is my desire, on the contrary, to add to my own debt what will be theirs also, and to remain so much the more,

Monsieur,

Your very humble and very obedient servant,

PASCAL

M. Perier received this letter at Moulins, where he was discharging duties which forbade him full freedom of action. Anxious though he was to carry out the experiment at once, he was not at liberty to do so before the month of September last.

The reasons for this delay, the story of the experiment, and the precision with which he conducted it, are made plain by the following letter with which he honored me.

Copy of the letter sent by Monsieur Perier to Monsieur Pascal the Younger, September 22, 1648

MONSIEUR,

AT LAST I have carried out the experiment you have so long wished for. I would have given you this satisfaction before now, but have been prevented both by the duties I have had to perform in Bourbonnais,[6] and by the fact that ever since my return the Puy de Dôme, where the experiment is to be made, has been so wrapped in snow and fog that even in this season, which here is the finest of the year, there was hardly a day when one could see its summit, which is usually in the clouds and sometimes above them even while the weather is clear in the plains. I was unable to adjust my own convenience to a favorable state of the weather before the 19th of this month. But my good fortune in performing the experiment on that day has amply repaid me for the slight vexation caused by so many unavoidable delays.

I send you herewith a complete and faithful account of it, in which you will find evidence of the painstaking care I bestowed upon the undertaking, which I thought it proper to carry out in the presence of a few men who are as learned as they are irreproachably honest, so that the sincerity of their testimony should leave no doubt as to the certainty of the experiment.

Copy of the Account of the Experiment submitted by Monsieur Perier

THE weather on Saturday last, the nineteenth of this month, was very unsettled. At about five o'clock in the morning, however, it seemed sufficiently clear; and since the summit of the Puy de Dôme was then visible, I decided to go there to make the attempt. To that end I notified several people of standing in this town of Clermont, who had asked me to let them know when I would make the ascent. Of this company some were clerics, others laymen. Among the clerics was the Very Revd. Father Bannier, one of the Minim Fathers of this city, who has on several occasions been "Corrector" (that is, Father Superior), and Monsieur Mosnier, Canon of the Cathedral Church of this city; among the laymen were

[6] A neighboring province.

Messieurs La Ville and Begon, councillors to the Court of Aids, and Monsieur La Porte, a doctor of medicine, practising here. All these men are very able, not only in the practice of their professions, but also in every field of intellectual interest. It was a delight to have them with me in this fine work.

On that day, therefore, at eight o'clock in the morning, we started off all together for the garden of the Minim Fathers, which is almost the lowest spot in the town, and there began the experiment in this manner.

First, I poured into a vessel six pounds of quicksilver which I had rectified during the three days preceding; and having taken glass tubes of the same size, each four feet long and hermetically sealed at one end but open at the other, I placed them in the same vessel and carried out with each of them the usual vacuum experiment. Then, having set them up side by side without lifting them out of the vessel, I found that the quicksilver left in each of them stood at the same level, which was twenty-six inches and three and a half lines[7] above the surface of the quicksilver in the vessel. I repeated this experiment twice at this same spot, in the same tubes, with the same quicksilver, and in the same vessel; and found in each case that the quicksilver in the two tubes stood at the same horizontal level, and at the same height as in the first trial.

That done, I fixed one of the tubes permanently in its vessel for continuous experiment. I marked on the glass the height of the quicksilver, and leaving that tube where it stood, I requested Revd. Father Chastin, one of the brothers of the house, a man as pious as he is capable, and one who reasons very well upon these matters, to be so good as to observe from time to time all day any changes that might occur. With the other tube and a portion of the same quicksilver, I then proceeded with all these gentlemen to the top of the Puy de

[7] This indicates a precision of measurement of less than a millimeter (1 *ligne* = 2.2. mm.).

Dôme, some 500 fathoms above the Convent. There, after I had made the same experiments in the same way that I had made them at the Minims, we found that there remained in the tube a height of only twenty-three inches and two lines of quicksilver; whereas in the same tube, at the Minims we had found a height of twenty-six inches and three and a half lines. Thus between the heights of the quicksilver in the two experiments there proved to be a difference of three inches one line and a half. We were so carried away with wonder and delight, and our surprise was so great that we wished, for our own satisfaction, to repeat the experiment. So I carried it out with the greatest care five times more at different points on the summit of the mountain, once in the shelter of the little chapel that stands there, once in the open, once shielded from the wind, once in the wind, once in fine weather, once in the rain and fog which visited us occasionally. Each time I most carefully rid the tube of air[8]; and in all these experiments we invariably found the same height of quicksilver. This was twenty-three inches and two lines, which yields the same discrepancy of three inches, one line and a half in comparison with the twenty-six inches, three lines and a half which had been found at the Minims. This satisfied us fully.

Later, on the way down at a spot called Lafon de l'Arbre, far above the Minims but much farther below the top of the mountain, I repeated the same experiment, still with the same tube, the same quicksilver, and the same vessel, and there found that the height of the quicksilver left in the tube was twenty-five inches. I repeated it a second time at the same spot; and Monsieur Mosnier, one of those previously mentioned, having the curiosity to perform it himself, then did so again, at the same spot. All these experiments yielded the same height of twenty-five inches, which is one inch, three lines and a half less than that which we had found at the Minims, and one inch and ten lines more than we had just found at the top

[8] *Ayant à chaque fois purgé très soigneusement d'Air le tuyau,* while filling and before inverting it.

of the Puy de Dôme. It increased our satisfaction not a little to observe in this way that the height of the quicksilver diminished with the altitude of the site.

On my return to the Minims I found that the [quicksilver in the] vessel I had left there in continuous operation was at the same height at which I had left it, that is, at twenty-six inches, three lines and a half; and the Revd. Father Chastin, who had remained there as observer, reported to us that no change had occurred during the whole day, although the weather had been very unsettled, now clear and still,[9] now rainy, now very foggy, and now windy.

Here I repeated the experiment with the tube I had carried to the Puy de Dôme, but in the vessel in which the tube used for the continuous experiment was standing. I found that the quicksilver was at the same level in both tubes and exactly at the height of twenty-six inches, three lines and a half, at which it had stood that morning in this same tube,[10] and as it had stood all day in the tube used for the continuous experiment.

I repeated it again a last time, not only in the same tube I had used on the Puy de Dôme, but also with the same quicksilver and in the same vessel that I had carried up the mountain; and again I found the quicksilver at the same height of twenty-six inches, three lines and a half which I had observed in the morning, and thus finally verified the certainty of our results.

The next day, the Very Revd. Father De la Mare, priest of the Oratory and Lecturer in Divinity of the Cathedral Church, who had witnessed all that had taken place on the morning before in the garden of the Minims, and to whom I had reported all that had occurred on the Puy de Dôme, proposed to me that I carry out the same experiment at the base and on the top of the highest tower of Notre Dame de Clermont, to see whether there would be any difference [between pressures

[9] *Serain.*

[10] *Ce mesme tuyau;* i.e. that first mentioned—the one which had been carried up the mountain.

at these heights]. To gratify the curiosity of a man of such great distinction who has given all France many proofs of his ability, I carried out that very day the ordinary experiment of the vacuum in a private residence which stands on the highest ground in the city, some six or seven fathoms above the garden of the Minims, and on a level with the base of the tower. There we found the quicksilver at the height of about twenty-six inches and three lines, which is less than that which was found at the Minims by about half a line. I next made the experiment on the top of the same tower, which was twenty fathoms higher than its base and about twenty-six or twenty-seven fathoms above the garden of the Minims. There I found the quicksilver at the height of about twenty-six inches and one line, that is, about two lines less than its height at the base of the tower, and about two and a half lines lower than it stood in the garden of the Minims.

Thus, collecting results and comparing the various elevations of the places where the experiments were made with the corresponding heights of quicksilver which were left in the tubes, it is found that:

In the experiment made at the lowest site the quicksilver stood at the height of twenty-six inches, three lines and a half; in that which was made at a place some seven fathoms above the lowest, the quicksilver stood at the height of twenty-six inches and three lines; in that which was made at a place about twenty-seven fathoms above the lowest, the quicksilver stood at twenty-six inches and one line; in that which was made at a place about one hundred and fifty fathoms above the lowest, the quicksilver stood at twenty-five inches; in that which was made at a place about five hundred fathoms above the lowest, the quicksilver stood at twenty-three inches and two lines.

Therefore, it is found that about seven fathoms of altitude give a difference in the height of the quicksilver of half a line; about twenty-seven fathoms give a difference of two lines and a half; about one hundred and fifty fathoms, fifteen lines and a half, which makes one inch, three lines and a half; and about five hundred fathoms, thirty-seven lines and a half, or three inches, one line and a half.

This is, in truth, all that took place in this investigation. All the gentlemen who assisted in it will subscribe to this account of it whenever you wish.

In addition, I must tell you that the heights of the quicksilver were read with very great accuracy; but those of the sites where the experiments were made, much less satisfactorily. If I had had sufficient leisure and adequate facilities, I would have measured the altitudes with more exactness, and would indeed have marked places on the mountain at intervals of a hundred fathoms, in each of which I would have made the experiment and recorded the differences in the height of the quicksilver which would have been observed, at each of these stations, in order to give you the exact difference caused[11] by the first hundred fathoms rise, that given by the second hundred fathoms, and so on, with the others. These data could serve for the construction of a table, by an amplification of which those who cared to take the trouble might possibly arrive at a perfect knowledge of the exact diameter of the total sphere of the air.[12]

I do not abandon the hope of sending you some day these hundred-fathom differences, as much for our own personal satisfaction as for the benefit the public might derive from them.

If you find any obscurities in this recital, I shall be able in a few days to clear them up in conversation with you, since I am about to take a little trip to Paris, when I shall assure you that I am,

Monsieur,

Your very humble and very affectionate servant,

PERIER

[11] *Au juste la difference qu'auroient produit les premieres cent toises . . .*

[12] This statement clearly implies that Perier suspected a diminution of the density of the air with increase of altitude; and his suggestion adds significantly to the evidence which the rest of his report provides of his thorough competence as an experimental investigator.

This narrative cleared up all my difficulties, and, I am free to say, afforded me great satisfaction. Seeing that a difference of twenty fathoms of altitude made a difference of two lines in the height of the quicksilver, and six or seven fathoms one of about half a line, facts which it was easy for me to verify in this city, I made the usual vacuum experiment on the top and at the base of the tower of Saint Jacques de la Boucherie, which is some twenty-four or twenty-five fathoms high, and found a difference of more than two lines in the height of the quicksilver. I then repeated it in a private house ninety steps high, and found a clearly perceptible difference of half a line. These results are in perfect agreement with those given in M. Perier's narrative.

Any who care to do so, may, for themselves, confirm them at their pleasure.

Consequences

FROM *this experiment many inferences may be drawn, such as:*

A method of ascertaining whether two places are at the same altitude, that is to say, equally distant from the center of the earth, or which of the two is the higher, however far apart they may be, even at antipodes—which would be impossible by any other means.

The unreliability of the thermometer in marking degrees of heat (which is not commonly recognized), as is shown by the fact that its liquid sometimes rises when the heat increases and sometimes, on the contrary, falls when the heat decreases, even though the thermometer be kept always in the same location.[13]

Proof of the inequality of the pressure of the air, which, at the same degree of heat, is always greater in the lowest places.

All these consequences will be fully set forth in the Treatise

[13] See p. 62, n. 7.

on the Vacuum, together with many others as useful as they are interesting.

TO THE READER

MY DEAR reader: The common people everywhere and the whole body of philosophers are agreed in asserting as a principle that nature would rather suffer her own destruction than tolerate the smallest empty space. Some of the best minds have taken a position less extreme than this, for although they have believed that nature abhors a vacuum, they have thought, nevertheless, that this repugnance has its limits, and that it can at times be overcome by force; but no one has yet been known to have advanced the third view, that nature has no repugnance for the vacuum, that she makes no effort to avoid it, and permits it without difficulty and without resistance.[14] The experiments set forth in my Abridgment[15] invalidated, in my judgment, the first of these principles; and I do not see how the second can withstand the evidence which I now present to you. Consequently, I now find no difficulty in accepting the third, namely, that nature has no repugnance to a vacuum, and makes no effort to avoid it; that all the effects ascribed to such abhorrence are due to the weight and pressure of the air, which is their only real cause; and that for lack of knowledge, people have purposely invented this imaginary abhorrence of the vacuum in order to account for them. This is far from being the only case in which, when the weakness of men has made them unable to discern true causes, their subtlety has substituted for them imaginary causes to which they have attached specious names which fill the ears, but not the mind. Thus they say that the sympathy and antipathy of natural bodies are the generic efficient causes of several effects, as if inanimate bodies were capable of sympathy and antipathy. It is the same with antiperistasis[16] and several other

[14] See p. 73, n. 5.

[15] The 1647 pamphlet; see p. xvii, n.

[16] The affinity of opposites, or blending of contrary qualities to produce that balance or harmonious adjustment which is the end toward which nature strives: a Greek idea which goes back to Heraclitus of Ephesus.

chimerical causes, which afford a vain alleviation to man's eagerness in his search for hidden truths, but which, far from disclosing these, serve but to cloak the ignorance of those who invent them and to feed the ignorance of their disciples.

Nevertheless, it is not without regret that I abjure these generally accepted beliefs. I only yield to the compulsion of truth. I opposed the new theories as long as I had any pretext for holding to the old: the principles I have used in my Abridgment sufficiently testify to this. In the end, however, the evidence of experiment compels me to lay aside the views which respect for antiquity induced me formerly to accept. I have departed from them indeed only little by little, and have discarded them by degrees, for from the first of these three principles—that nature has an unconquerable abhorrence of a vacuum, I passed on to the second—that she does feel that abhorrence, but not insuperably; and thence, at last, have come to believe the third—that nature has no such abhorrence at all.

This is the position to which I was brought by this last experiment on the equilibrium of fluids, which would have been incompletely laid before you had I not shown you the motives that urged me to perform it. It is for this reason that I include my letter of the sixteenth of last November addressed to M. Perier,[17] who has taken great pains to make the tests with all the care and precision that could be desired, and who is fully entitled to the gratitude of all the interested men who have so long looked forward to this investigation. And since by happy chance their universal desire had made it famous even before it was made known, I am confident that it will be no less so after its performance, but will afford as much satisfaction as its expectation has caused impatience.

It would not have been proper to keep waiting any longer those who wish to be informed concerning it; and it is for this reason that I have not been able to restrain myself from issuing an account of it in advance, in spite of my first intention to withhold it for publication in the complete Treatise

[17] See pp. 98 ff.

(which I have promised you in the Abridgment), where the consequences I deduced from it will be set forth. I had planned to defer the completion of this work until this last experiment was made, because it was necessary for the verification of my demonstrations. But since the Treatise cannot appear at this early date, I wish to withhold the narrative no longer—as much that I may deserve your greater gratitude by my promptness, as that I may avoid the reproach of wronging you, as I think I should, by further delay.

RECORD

OF THE OBSERVATIONS taken by Monsieur Perier continuously day by day, during the years 1649, 1650, and 1651 in the city of Clermont in Auvergne, on the variations in the rise and fall of quicksilver in tubes; and also of those made simultaneously of the same variations in Paris by one of his friends, and in Stockholm in Sweden, by Messieurs Chanut and Descartes.

AFTER my experiment at the Puy de Dôme, an account of which is given above, Monsieur Pascal wrote to me from Paris to Clermont where I was at the time, that not only differences of location, but also differences of weather in one location, according as it was more or less hot or cold, dry or humid, would cause different elevations or depressions of the quicksilver in the tubes.

In order to find out whether this were true and [if so] to ascertain whether changes in the state of the weather caused this variability with sufficient constancy and regularity [of correspondence] to make the establishment of a general rule possible and the determination of its cause unequivocal, I resolved to make several measurements of these effects during a considerable period of time.

The more easily to conduct them, I set up a tube with its quicksilver for continuous operation, fastened it in a corner of my study, and ruled it in inches and lines from the surface of the quicksilver where it dipped, to the height of thirty inches. I read it several times a day, but especially at evening and morning, and recorded on a sheet of paper the exact height at which the quicksilver stood every morning, every evening, and even sometimes at midday, whenever I noticed differences at that time. And I also noted the [corresponding] differences of weather, to see whether the one effect always followed the other.

I started these observations at the beginning of the year 1649, and continued them until the last of March in 1651. After carrying them on for five or six months I had noticed large fluctuations in the height of the quicksilver. I found indeed that ordinarily, and as a usual thing, the quicksilver did, as I had been informed, rise in the tubes during cold, humid, or cloudy weather, and fall when the weather was hot and dry. But such was not invariably the case. It happened at times that, on the contrary, the quicksilver fell when the weather became colder or damper and rose when it turned more warm or dry.

To improve and to increase my knowledge of these correspondences, it occurred to me to try to obtain simultaneous readings made in other places far distant from one another, in order to see whether by the comparison of these results I could discover anything further. With that in view I broached the subject to a friend of mine, a man of great thoroughness in all things, who was at that time in Paris. I requested him to make the same observations in Paris that I was making at Clermont and to send me his records for every month. He did so from the first of August, 1649, to the end of March, 1651, at which time I myself concluded my observations.

I took the liberty also to submit the matter by letter to M. Chanut, whose merit and reputation are known throughout Europe, who was at that time ambassador to Sweden. He did me the favor to grant my request and likewise to send me the observations that he and Monsieur Descartes made in Stockholm from October, 1649, to September, 1650. In exchange, also, I sent him my own.

But from the comparison of all these observations I could derive no further advantage than the confirmation of my own results, which showed that ordinarily and usually the quicksilver rises in cold, cloudy, and damp weather, and that it falls in hot and dry weather, or when rain or snow is falling; but that this does not always happen, and that sometimes the very opposite is the case, when the quicksilver rises as the weather becomes warmer and falls when it grows colder, or even falls when it turns cloudy and damp, and rises when there is more rain and snow; so that no general rule can be formulated.

This much, however, I think we might lay down with some degree of certainty—that the quicksilver rises whenever these two things happen together: that the weather turns cooler and also becomes highly charged and clouds over; and that on the contrary the quicksilver falls whenever these two things happen together: that the weather turns warmer and also discharges [its moisture] by rain or snow. But whenever only one of these two things occurs, for example when the weather merely turns colder without clouding over, it may well be that the quicksilver does not rise, although ordinarily the cold makes it rise, because the air happens to be in some condition[1] such as [that occasioned by] rain or snow, which produces a contrary effect. In this case, that one of the two conditions which prevails, either the cold or the snow, determines the effect.

M. Chanut had conjectured from his observations during the first twenty-two days, that it was the prevailing winds which caused these various changes. But it does not seem to me that this conjecture can stand against the following experiments. Indeed Monsieur Chanut, as his letters show, had himself clearly foreseen that they might invalidate it. As a matter of fact the quicksilver both rises and falls in all sorts of wind and in all seasons of the year, although it is usually higher in winter than in summer. I say ordinarily, because the rule does not always hold. For instance, at Clermont on January 16, 1651, I saw the mercury standing at twenty-five inches

[1] *Parce qu'il se rencontre une qualité en l'Air . . .*

and eleven lines, and on the 17th at twenty-five inches and ten lines, which is almost its lowest level. On one of these days the weather was mild and still, on the other there was a strong west wind. In Paris on August 9, 1649, it was observed at twenty-eight inches and two lines, a height which it almost never exceeds. I cannot say what the weather was there, since the observer at Paris made no record of it. Nevertheless the following general remarks may be made concerning the highest and the lowest levels observed in the following experiments.

At Clermont: highest 26 inches, 11 lines and a half; February 14, 1651. Ice abundant in north; weather fairly clear. That did not happen excepting on this one day; but on many other days during that same winter we found 26 inches, 10 lines, or 9 lines, and even 11 lines on November 5, 1649.

Lowest, 25 inches, 8 lines, October 5, 1649. No other registrations were as low as this; a few others were 25 inches, 9 or 10 or 11 lines.

The difference between the highest and lowest registrations at Clermont is 1 inch, 3 lines and a half.

At Paris: the highest, 28 inches and 7 lines, November 3 and 5, 1649; the lowest, 27 inches, 3 lines and a half, October 4, 1649.

It may be noted that in the same month of this year the quicksilver stood almost at its highest and lowest, namely, at 28 inches, 6 lines on the 4th, and at 27 inches and 4 lines on the 14th of December, 1649.

The difference between the highest and lowest in Paris is 1 inch, 3 lines and a half.

At Stockholm: the highest, 28 inches and 7 lines, was reached on December 8,. 1649, on which day Monsieur Descartes reported severe cold. The lowest was 26 inches, 4 lines and three-quarters on May 6, 1650; wind southwest, weather unsettled and mild.

The difference between highest and lowest in Stockholm is

two inches, 2 lines and a quarter. Thus the inequalities proved far greater at Stockholm than at Paris or Clermont.

These inequalities are at times very sudden. For instance, on December 6, 1649, 27 inches and 5 lines, and on the 8th of the same month 28 inches and 7 lines.

I might easily have had most of these observations printed, as I still have the original records; but I thought few people would be interested in them. Nevertheless, it could yet be done, if it were desired. Meanwhile I add here two letters from Monsieur Chanut, which I have mentioned before, and which confirm all I have said of him in this narrative.

Copy of a letter written by Monsieur Chanut to Monsieur Perier from Stockholm, March 28, 1650

MONSIEUR,

A VERY few days after I wrote you the letter which you were kind enough to answer on the eleventh of March last, we lost Monsieur Descartes from a malady similar to that from which I had suffered a very short time before. As I write I am still grieving for his loss; for even his learning and his intellect were less than his greatness, his goodness, and the innocence of his life. His servant, on leaving his dwelling, has not remembered to leave with me the record of his observations on quicksilver, like that he sent to you. But when I received your copy, my interest was aroused afresh, and I decided that, in glancing once a day at a corner of my study, I would not be seriously neglecting my duty in service to the King. I therefore began observing on the sixth of this month; and considering that if what you wrote me was correct, all our observations would be fruitless, I was not willing to accept the conclusion indicated by your experiment, namely, that the temperature and movement of the air cause no regular changes.

So I added to my own observations of heat and cold, dryness and humidity, cloudiness and fair skies, that of the prevailing winds, which, it appears, Monsieur Descartes had not observed.

Now I have found, after twenty-two days of observation in diversified and changing weather—this season is always uncertain in these parts—that the prevailing winds cause uniform and almost regular increase or diminution of the mercury in its tube. I cannot believe that this fact escaped such keen observers as yourselves, and should rather be inclined to think you were trying to provide M. Descartes with a problem when you withheld this peculiarity from him. I shall continue until I grow weary of the tests and, if you wish, shall send you a copy of my daily record, which will show you exactly what happened in my study. I also beg you to supply me with an account of your observations, without omitting your notes on the winds; for, here, it is in these that I find the ever-present cause of variation in the height of the mercury in the tube. Possibly the following experiments may invalidate this first conjecture of mine, which I submit to you without claiming to tell you anything new. I sincerely hope that your brother-in-law, Monsieur Pascal, who has the necessary leisure and a very remarkable penetration, may find in this matter some indications of importance to Physics. I should be happy if our Northland should yield him some observations to aid him in his quest. And I shall value them the more for affording me the opportunity of assuring you more frequently that I am,

MONSIEUR,

Your very humble and obedient servant,

CHANUT

Copy of another letter from the same Monsieur Chanut to the same Monsieur Perier, Stockholm, September 24, 1650

MONSIEUR,

I RECEIVED, with the letter you did me the favor to write on July 29, your memorandum of observations, which I value highly as a mark of the good will with which you honor me and as a source of delightful study a little later when I shall enjoy greater leisure than my public duties leave me at present. I beg of you patience until then. I think constantly of continuing my observations, which shall be food for discussion some day when the opportunity offers. Meanwhile, as a small return for the trouble you have taken to write to me, I must tell you that the late Monsieur Descartes had intended to continue these same observations in a glass tube provided near the middle with a container or large bulb at about the height to which the quicksilver almost rises. He would have let water [into this bulb] above the quicksilver, until it reached about half way up the remaining height,[2] and would thus [on its larger surface] have observed the changes more distinctly. I wished to try that method; but our glassblowers are unskillful, and have no place suitable for the annealing of tubes with that container or bulb in the middle. All their tubes cracked; and consequently I have no data at hand other than of the usual sort; but these I send you, for what they are worth. If this discussion, which you were so good as to think well of, does not improve our knowledge of nature, at least it will serve, if you are willing, to maintain our friendship. In addition, I beg you to grant the favor of encouraging my friendship with the Messieurs Pascal. My wife and I present

[2] . . . un tuyau de verre, vers le milieu duquel il y eût une retraite où un gros ventre, environ à la hauteur ou monte à peu près le vif argent, au dessus duquel vif argent mettant de l'eau jusqu'au milieu environ de la hauteur qui reste au dessus du vif argent; . . .

our very humble respects to Madam Perier and to Mademoiselle Pascal; and we are not without hope that some day we may have the pleasure of greeting you in the Province. I am,

MONSIEUR,

Your very humble and very obedient servant,

CHANUT

And the superscription to both letters was:

To M. Perier, Councillor to the King in the Court of Aids of Clermont-Ferrand,

AT CLERMONT

NEW EXPERIMENTS,

Made in England, explained by the principles set forth in the two foregoing Treatises on the Equilibrium of Liquids and the Weight of the Mass of the Air.

IN ADDITION to the experiments which have been described in the foregoing treatises, a great number of others of the same kind may be made, all of which may be explained by the principle of the Weight of the Mass of the Air.

Within the last fifteen or twenty years several men have enjoyed devising new experiments of this sort. And among others, an English gentleman named Monsieur Boyle has made some that are very curious which can be seen in a book that he has written in English and since translated into Latin under the title *Nova Experimenta Physico-Mechanica de Aëre.*[3]

We have thought it appropriate to set down here, in brief, the most significant of them, to show their relation to those described in the foregoing treatises, and to confirm yet further the principle there established of the weight of the mass of the air.

[3] The original work appeared at Oxford in 1660 under the title *New Experiments Physico-Mechanical Touching the Spring of the Air and Its Effects.* It is now somewhat scarce; but is generally accessible in *The Works of the Honourable Robert Boyle*, edited by Thomas Birch (London 1744-72, pp. 1-117). Boyle's discovery of the inverse proportionality between the volumes and pressures of masses of air at constant temperature—Boyle's Law—was announced in a second discourse which accompanied the second edition of 1662 and was entitled *A Defense of the Doctrine Touching the Spring and Weight of the Air . . . against the Objections of Franciscus Linus.* Birch edition of the *Works,* I, 118-85. A third English edition appeared in 1682.

One of the most remarkable features of this book of experiments by Monsieur Boyle is the account of the machine he used in performing them. Since it is impossible to withdraw all the air from a chamber, the only way of producing a vacuum that had thus far been thought of was to empty the upper closed end of a tube by mercury; and this vacuous space was so small that with it no extensive experiment could be carried out.

But by the use of a machine which was first invented by certain philosophers of Magdeburg,[4] but which he himself[5] later greatly improved, M. Boyle has found means to evacuate a very large glass vessel with a big opening at the top through which anything one pleases can be introduced; while the transparency of the glass makes visible what happens when the vacuum is made.[6]

The machine consists of two principal parts, namely, a large glass vessel which he calls the receiver because of its resemblance to the vessels used by chemists and so called by them, and another vessel which he calls the pump, because it serves to draw out and suck up the air contained in the receiver.

The first of these vessels, that called the receiver, is shaped

[4] *Dont la premiere invention est deüe à ceux de Magdebourg* . . . Reference is here made to the pioneering work of Otto von Guericke, who, in 1655 or thereabout, first produced sizable vacua by pumping the air out of large copper vessels, and who also, it must be recognized, anticipated many of Boyle's most striking experiments. (See the adequate summary of the evidence given by Rosenberger: *Geschichte der Physik* II, Braunschweig, 1884, pp. 144-50, from Hoffmann: *Otto von Guericke,* Magdeburg, 1874.) Von Guericke described his completed work on vacua in a book published in 1672—which also contained an account of his equally original experiments on static electricity, together with certain cosmological speculations—entitled *Experimenta nova, ut vocantur Magdeburgica, de vacuo spatio.* Before this time it had become known only through the account of it given by Kaspar Schott in his *Mechanica hydraulica,* published in 1657, which was known to Boyle. See the dedicatory letter in *New Experiments.* . . .

[5] With the probably very effective assistance of Robert Hooke, which Boyle acknowledged.

[6] The following is a careful and perfectly correct condensation of Boyle's own description in his *New Experiments,* which is elaborately minute. Since it suffers somewhat for lack of a drawing, a copy of the original (*New Experiments,* Plate. Reproduced in the Birch edition Vol. I, Plate I) is here included. The designations "receiver" and "stopcock" are Boyle's own.

round like a ball, so as to be stronger and better able to withstand the pressure of the air when it is being exhausted; and it is large enough to hold sixty pounds of water at sixteen ounces to the pound, that is to say about thirty pints in Paris measure. This, he says, is the largest one that glass workers could blow.

Its very large opening at the top has a lid suitable for closing it; and this itself has an aperture in its center which may be closed by a stopcock which one raises[7] more or less or removes entirely in order to allow as much air as he wishes to flow back into the receiver after it has been evacuated. In addition to this upper aperture the receiver has also another one at the bottom which is somewhat tapered, and fitted to one of the openings of another stopcock.

The other part of the machine, called the pump, is made of brass[8] in the form of a hollow cylinder some thirteen or fourteen inches long, with an inside diameter of about three inches. It has two openings at the top, one of which receives the lower outlet of the stopcock, of which the other end, as we have said, enters the bottom of the receiver; so that by this means there is a communication between the receiver and the pump when the stopcock is opened. The other opening, close to the first, is one by which the air can be let out from the pump or hollow cylinder; and on this is a valve that allows air to escape from within but prevents the air outside from entering.

The pump is altogether open at the bottom, but may be closed by a large piston, which is accurately fitted so that air cannot pass between it and its casing.

This piston has for its handle a narrow but substantial bar of iron, somewhat longer than the cylinder, one side of which, cut along its whole length, is toothed: that is, covered with indentations into which there enter the teeth of a wheel which is fixed to a wooden frame that serves to support the cylinder and the rest of the machinery. By revolving this wheel one

[7] That is, this upper cock is a stopper, not a turncock.
[8] *Airin.*

makes the piston move up or down at will, and in this way [sucks air from the receiver and] drives out the air contained in the cylinder. This air escapes by the upper aperture, which is then immediately stoppered by a special brass[9] plug nicely turned to fit it exactly.

This description suffices for the understanding of the experiments about to be described. Those who wish to see a more complete and particular account can find it in M. Boyle's book, which contains a plate engraving of the machine.[10]

To empty the receiver by means of this machine, the piston, at the start, must be at the bottom of the cylinder, the stopcock guarding the passage from the receiver to the pump must be closed, and the aperture at the top of the cylinder open.

These things being so arranged, the piston must be raised by means of the wheel to the top of the cylinder and thus made to drive out all the air it contains through the aperture at the top, then open, which immediately thereafter is plugged with the brass stopper. Next, the piston must be brought down again to the bottom of the pump, so that by this means it is completely emptied of air. After this, the stopcock which makes connection between the receiver and the pump must be opened; then the air from the receiver, streaming through the stopcock, fills the pump, which must be emptied again, as before, after closing the faucet. The process of filling and emptying is repeated till no air is heard escaping through the aperture at the top of the pump, and a lighted candle held close to it is no longer blown out—which proves that no air is now being drawn from the receiver and that as complete a vacuum exists within it as can be made by this machine.

It is easy to understand, however, that a perfect vacuum cannot be effected by this means, as Monsieur Boyle himself confesses; because when the pump has been emptied and the stopcock opened, the air of the receiver does not all enter the pump, but is distributed between the two vessels in proportion

[9] *Cuivre.*

[10] Here reproduced.

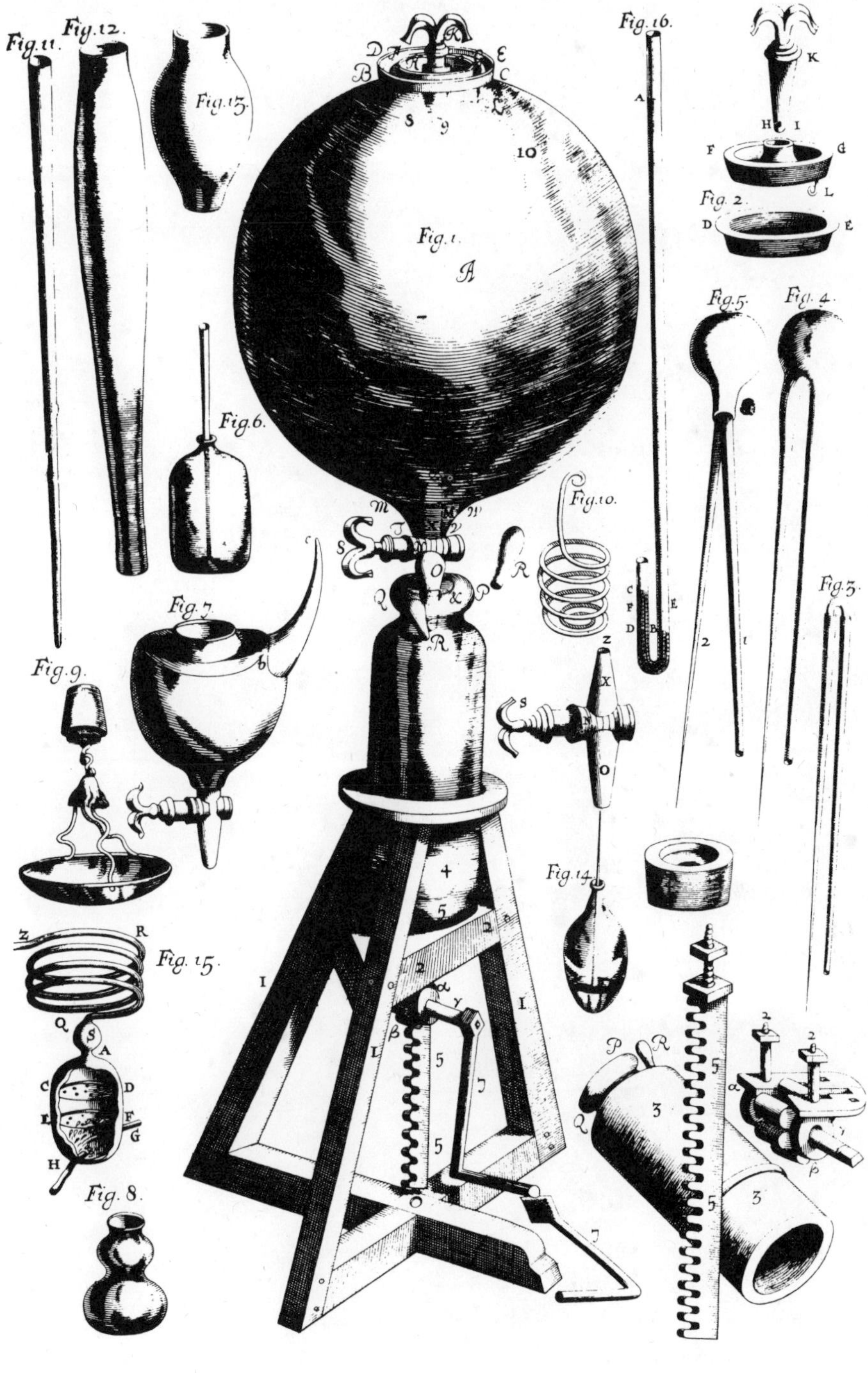

Fig.1.
Fig. 2.
Fig.3.
Fig. 4.
Fig.5.
Fig.6.
Fig.7.
Fig. 8.
Fig.9.
Fig.10.
Fig.11.
Fig.12.
Fig.13.
Fig.14.
Fig. 15.
Fig.16.

to their capacities; and since the receiver is much larger than the pump, there remains a greater quantity of it in the former than in the latter. Nothing could prevent a slight amount from being left in the receiver unless the capacity of the pump were incomparably greater than that of the receiver, which was not the case. It is, therefore, not to be wondered at that some of the effects are not what they ought to be if there were a complete vacuum. For instance, the quicksilver does not drop all the way down in the ordinary experiment, and similarly when water is used, it remains suspended at a considerable height.

But this much should be observed, that if the anticipated effects are not produced in every particular, they are, nevertheless, in great part produced, and proportionally to the amount of air sucked from the receiver. For instance, as Monsieur Boyle states in his account of the experiment, the quicksilver does not remain suspended at the height of twenty-seven inches as it would in the air, but only at the height of one finger's breadth, that is to say, at nine or ten lines, while the water does not remain suspended at the height of thirty-two feet, but at that of one foot only, which is a change proportional to that shown by the quicksilver. This is a great diminution, which shows that these effects are due to the weight of the air, of which only a small amount is left in the receiver; and quite as clearly as if the water and the quicksilver dropped to the very bottom in a complete vacuum.

Certainly nothing could show more clearly that it is the weight of the mass of the air that produces all these effects, which one also observes in the liquids which remain suspended, some higher, some lower, in the ordinary vacuum experiments. By nothing else is this better shown than by the fact that these effects cease entirely when the pressure and resiliency of the air are removed, as happens when the experiment of a vacuum within a vacuum is made.[11] These effects also diminish very perceptibly and are almost reduced to nothing when the air

[11] Pages 99 f.

which presses the vessel in which the liquid is spread out[12] is reduced to a minimum, as it is in this machine of Monsieur Boyle.

And that is why, although some experiments may be made with that receiver which seem quite similar to those in the open air (as, for instance, when two polished bodies remain attached together in it without dropping apart after the pump has exhausted the air), it does not follow that this effect can be produced as well in a vacuum as in the air, and therefore is not caused by the weight of the air. This would be contrary to what has been stated in the Treatise on the Weight of the Mass of the Air. It follows only that this effect is due to the air which has remained in the receiver, which, though it expands and becomes rarer because it is no longer compressed by the outer air and by its resilience presses together the two bodies, still has strength enough to prevent them from falling apart. Since, however, they are not then subjected to as great a pressure as they are in the outer air, one would doubtless not feel as great a resistance to their separation if it were possible to put the hands into the receiver. An easier way to make the experiment would be to hang on the lower body a fairly heavy weight that would have the same effect as a hand pulling it down: one would then observe that after the receptacle was exhausted the two bodies would separate much more readily than they do in the air.

Thus this experiment is quite similar to those we have spoken of made with water and quicksilver by means of this machine, as we have related. For if in the experiment with water [suspended in a barometric tube], in which the water empties itself to the level of one foot, one used instead of a tube three or four feet long another only half a foot in length, the effect would be that when the receiver was emptied of air the water would not fall at all, but would remain permanently suspended at the top of the tube, because the remaining air would still be equal to holding it up to that height. And just

[12] That is, which presses the liquid surface in the basin of the barometer.

as one may not infer from this that water would remain similarly suspended in taller tubes, say three or four or any number of feet high, and that consequently this suspension of the water is not due to the pressure of air; so also one may not infer from the fact that two bodies weighing say four or five ounces or slightly more each, adhere to one another in that receiver, that two much heavier bodies will similarly adhere, and that therefore this adhesion of two polished bodies in close contact is not due to the weight of the air.

Thus it is clear that, among all the experiments that can be made with this machine, the particular tests in which effects occur like those we have just discussed do not militate against the principle of the weight of the air, since it may justly be claimed that they are due to the air that is left in the receiver; while the other experiments, on the contrary, help to prove and establish the principle just as conclusively as if the receiver had been wholly emptied.

We shall now, therefore, describe a few experiments, taken, as we have said, from Monsieur Boyle's book, and show that they are manifestly dependent upon the principle of the weight of the air.

I. He notices, first of all, that after exhausting the receiver as we have described, it is quite difficult to lift the plug of the stopcock[13] at the top of the receiver, which feels as heavy as if a big weight were hanging from its lower end.

It is very natural and easy to explain this by the principle of the weight of the air; for in this experiment the air does not touch the plug underneath, but only above; and to lift it, the column of air which presses upon it must be lifted. Since this column is heavy, it is not astonishing that the key is felt to be heavy and hard to raise.

II. He notices also that after having forced the piston to the top of the cylinder, thus expelling all air, there is much difficulty in bringing it down again: it seems stuck fast to the

[13] *La clef de robinet.*

top of the cylinder so that considerable force must be applied to pull it away.

This effect is no harder to explain than the preceding. Since the air which surrounds the piston presses it below and not above, it is necessary to push away and lift the column of air that presses against the bottom in order to bring it down; and this can be done only with difficulty and by the application of considerable force.

III. He then describes several experiments that he made within the receiver. First he used a lamb's bladder, rather large, dry, quite soft, and only half-filled with air. Having closed its mouth tightly so that no air could get in, he put it while in that condition into the receiver which, after its aperture had been tightly closed, he exhausted by means of the pump. As fast as the receiver was emptied, the bladder was seen to swell, until, even before the receiver was as well exhausted as it could be, the bladder seemed completely expanded and as taut as if newly inflated. Then, in order to assure himself that the inflation of this bladder was due to the removal of the air which had surrounded it and pressed upon it, he raised the plug of the stopcock at the top of the receiver very slightly, so as to let air in gradually; and as fast as the air entered, the bladder was seen to relax little by little until finally on the free return of air, it became as flabby as it was before. He relates in this connection an experiment quite similar, made with a carp's bladder, which he says was devised by Monsieur de Roberval.[14] The experiment with the lamb's bladder he repeated several times and remarked that when he left too much air in it, it burst, with a noise like that of a petard.

To account for this effect in accordance with our principle, a word will suffice. It is exactly like the one that was described in the Treatise on the Weight of the Air (page 31), of a

[14] Gilles Personne de Roberval (1602-75). French mathematician, a friend of Mersenne and Pascal, who participated in the lively discussions of their circle on matters mechanical; distinguished, however, more for his contributions to the early development of the methods of the infinitesimal calculus than for his work in physics.

balloon which is inflated or deflated when it is taken up to the top of a mountain or brought down again; for the lamb's bladder is seen similarly to swell as the air which compressed it and made it look flabby and shriveled diminishes.

IV. He notes also, by the evidence of several actual experiments, that in exhausting a glass vessel which is not round but oval in shape, it invariably breaks, be the glass never so thick; while if it is altogether round like a ball, even though it is very much thinner, it does not break: because the spherical shape causes the various parts mutually to support and strengthen one another.

This effect is not due to the abhorrence of nature for a vacuum, for if that were the case the spherical vessel ought also to break like the other. It is caused by the weight of the air, which, pressing heavily upon the two vessels from the outside and very little from the inside, since they are almost exhausted, breaks the one which is oval in form because it has less resistance, but not the round one because its shape makes it stronger and better able to withstand the effort the air makes to break it.

V. It is also by this same principle of the weight of the air that we must explain the result of another experiment he describes, made with a water-filled siphon, eighteen inches long, which he placed inside his receiver and which ceased to flow as soon as, by means of the pump, he had emptied that receiver of air. It is clear that, since the air which remains in the receiver can by its pressure raise the water only to the height of one foot, as stated above, an eighteen-inch siphon must cease to flow.

VI. He also proved that bodies of different sizes which weighed the same in the air, lost their equilibrium in a vacuum. The experiment was carried out as follows: He took a dry bladder, half full of air, closed its aperture tightly and then tied it to one arm of a beam balance so accurate and so sensitive that the thirty-second part of one grain could depress it either way. To the other arm of the balance he attached an amount of lead of the same weight as the bladder. The two weights were thus balanced in the air: indeed he noticed that the lead was slightly heavier than the bladder. But when the whole device was set in the receiver and the air exhausted by the pump, it was seen, on the contrary, that the scale pan from which the bladder hung weighed more than the other, and fell further and further as the air in the receiver was sucked out; while when the air was gradually let in again, the bladder was seen to rise once more, little by little, regaining its full equilibrium when the air was allowed to re-enter freely.

This effect is precisely the same as that which has been noted in the Treatise on the Equilibrium of Liquids (pp. 17, 18): weights balanced in the air may not be balanced in water, nor even in a damper atmosphere. The reason we gave in the passage referred to serves also to explain the experiment here described. For it is clear that when the bladder balances the lead in the open air it is then counterpoised not only by the lead but also by a volume of air equal to itself and much greater than that which counterpoises the lead. Now when the balance is set up in this almost empty receiver, then although the natural weight of the bladder is not increased, it is nevertheless less counterpoised and supported because the volume of the air which upholds it has lost much of its force by the diminution of the total mass of air, and much more in proportion than the air that counterpoises the lead, because it is much greater. Hence the bladder, which balanced the lead in the air, must fall in the vacuum and cease to be in equilibrium.

In addition to these experiments Monsieur Boyle made a few others, which do not at all depend upon the principle of the weight of the air and would yield the same results even if the air had no weight, but which, nevertheless, are not inconsistent with this principle.

He has ascertained, for instance, that a pendulum does not swing as fast in the air as it does in a vacuum. In order to realize this, he took two pendulums that were perfectly equal in the air,[15] and of these put one only in the receiver, leaving the other in the air. When the receiver was exhausted, the pendulum within it swung faster than the other in the open air: in such degree that twenty-two beats of the former equaled only twenty beats of the latter.

He also noticed that sounds lost much of their intensity in the receiver when it was exhausted. This he proved by means of a repeating watch which he put in this receiver. The striking of the watch could hardly be heard after the receiver was exhausted, although it had been plainly audible before.

This does not, to my thinking, contradict what we have said about the experiment with the bladder that burst with the loud noise of a petard; because all that can be correctly inferred from the comparison is that the latter sound must have been much louder.

He wished also to ascertain whether fire could burn in the exhausted receiver, and how long it could last. For this purpose, he first put there a lighted tallow candle, which, he reports, went out in less than one minute after the receiver had been exhausted of air. When he made the same experiment with a small taper of white wax, it likewise did not remain alight for more than one minute. Next he put in glowing embers, and having very soon emptied the receiver of air, he noticed that, from the time he began to empty it to the moment when the live coals were quite extinguished, only three minutes had elapsed. Having in the same manner put in red-hot iron instead of coals, its redness remained visible for a space of four minutes. Again he made the same test with a small length of the slow match used by soldiers for their muskets. He hung it, lighted, in his receiver, and it went out like the rest as fast as the air was sucked away.

[15] *Parfaitement égaux.* This is correct. Boyle says (Experiment XXVI) that the bobs were of steel, equal in weight and as close to the same size as the mechanic could make them. He hung one in the receiver and one outside; and before exhausting, adjusted the length of the thread by which the latter was suspended until it swung in phase with the other. Then, in the test made after exhaustion, he adjusted the amplitude of its swing to equal, as closely as he could see, that of the one inside.

He wished next to find out what would happen to animals that were put into the receiver: whether the winged ones would fly, the others crawl, and finally whether either the one sort or the other could live there for any length of time. He first put in the winged creatures, such as blue-bottle flies, bees, and butterflies. After the receiver was emptied of air they fell straight down, unable to use their wings. He next put in a lark, which not only lost the use of its wings but became suddenly languid and after several violent convulsions was seen to die. All this took place in some nine or ten minutes. A sparrow, next put in, likewise died in five or six minutes. A mouse survived slightly longer, and suffered fewer convulsions than the winged creatures.

Wishing to ascertain whether fishes could live under those conditions, and unable to procure any other live ones, he put in an eel, which, after he had emptied the receiver, lay motionless on the bottom for a long time, as if dead. Nevertheless, when the receiver was opened and it was lifted out, he found that it was not dead, but on the contrary as lively as when it was first put in.

This much it has seemed appropriate to select from Monsieur Boyle's book: it includes the experiments we found to be most significant and those which bear most closely on the matter of the preceding treatises. Among them some are of particular importance in that they clearly prove that the air has weight; and all of them have this in common: that they prove nothing contrary to that principle.

APPENDICES

Appendix I

SIMON STEVIN

FOURTH BOOK OF STATICS

The Elements of Hydrostatics

Translated by Ada Barry from *Les Œuvres mathematiques* (1634)[1]

Argument

FIRST *we shall set forth definitions of the terms proper to this art, together with the necessary postulates*[2]*; then the propositions. Of these, the first nine will state some of the properties of bodies in water. The tenth, eleventh, twelfth, thirteenth, fourteenth, and fifteenth propositions will treat of the force with which water presses against its bases or retaining surfaces.*[3] *The sixteenth and seventeenth propositions will demonstrate the lengths of sides necessary in order to have any desired pressure of water against those surfaces. The eighteenth, nineteenth, and twentieth propositions will concern the centers of gravity of the total pressures of water*[4] *against the bases or retaining surfaces. The twenty-first proposition, will show how to find* [*the volume of*] *a quantity of water by its weight. The twenty-second and last proposition will deal*

[1] *Les Œuvres mathematiques de Simon Stevin de Bruges—augmenté par Albert Girard,* Leiden, 1634. Second part (Vols. II-VI), pp. 484 ff.

[2] *Petitions* (petitiones) in the legal or logical sense of claim or assertion.

[3] *Fonds ou retinacles* (retinacula); see Definition VIII, below.

[4] *Centres de gravité des pressemens de l'eau:* points of average pressure on the surfaces.

with the proportion that obtains between the sizes of solids, the densities of their matter[5] *and their weights. And finally, there will follow an Appendix on the practice of hydrostatics.*

DEFINITIONS

DEFINITION I

A known weight[6] *is that of which the known quantity is expressed in terms of known masses.*[7]

DEFINITION II

Parigrave substances are those of which equal volumes[8] *are also equiponderant.*

DEFINITION III

Multigrave substance is that which in comparison with others of the same volume is nevertheless heavier.

DEFINITION IV

Minugrave substance is that which in comparison with others of the same volume is lighter.

DEFINITION V

And, of bodies of equal size, as many times as the multigrave is heavier than the other, so many times denser[9] *is it said to be than the other, as dupligrave, tripligrave, etc.*

DEFINITION VI

A solid body is that which is not liquid, nor fluid, and which does not liquify in the liquid in question.

[5] *Pesanteurs de matiere.*

[6] *Pesanteur cognue.*

[7] *Poids.*

[8] *Grandeurs.*

[9] *Plus pesante.* The artificial words *multigrave,* etc., are obviously used to give unambiguous definition to the conception of density. The word *pesante,* as commonly used by Stevin himself, may mean either heavy or dense, and *pesanteur,* either weight or density.

DEFINITION VII

Vasiforme[10] *is that which is only the exterior surface of the body which it contains; and from which it may be separated in imagination.*

DEFINITION VIII

Fond is any surface against which water rests.

DEFINITION IX

Fond convenant is that surface of which every two halves are exactly alike. It might be said to be that surface in which all the diameters are cut in two equally by the center.

DECLARATION

The circles, ellipses, parallelograms, and regular polygons with sides of even number, and all other figures of whatsoever mixings of lines that can be made, such as A and B, which are cut in two equally by the straight line which passes through the center are called *fonds convenans* to distinguish them from others, regular and irregular, which are not so cut in two equally by the straight lines passing through their centers,

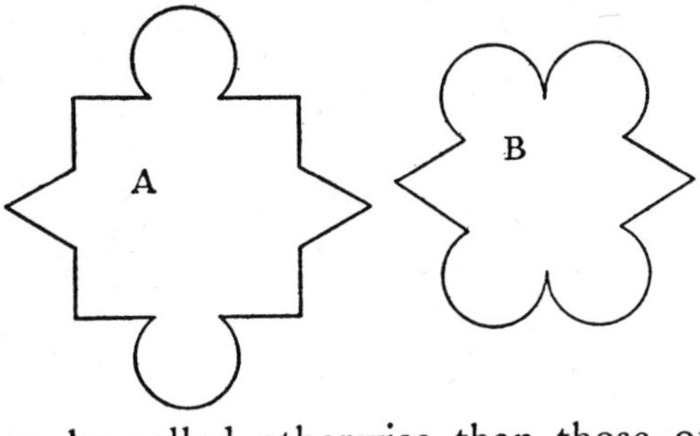

and which will therefore be called otherwise than those of this definition, namely, *fonds inconvenans,* such as triangles and polygons with sides of uneven number. The reason for this definition (as will be seen later) is that the columns having such bases, being cut by a diagonal plane passing through two homologous points in top and base, are cut in two equally.

[10] Another artificial word which designates succinctly a conception essential to Stevin's manner of proof, and very frequently used.

DEFINITION X

Vuide is a place in which there is no body.[11]

DEFINITION XI

Vuidé is a vessel inside which there is nothing but air.

Postulates

POSTULATE I

Let the true weight of a body be that which is found when it is weighed in the air; but in water, let the weight be called the state or condition[12] *of the body in this medium.*

POSTULATE II

Let the water designated be of uniform density.

POSTULATE III

Let the weights which do not sink as far as others be said to be lighter, but those which sink further, heavier; and those which sink equally, equiponderant.

POSTULATE IV

Let it be assumed that the vasiforme may contain water or other matter without breaking or changing its shape.

POSTULATE V

Let it be possible for the vasiforme full of water to remain emptied of all matter but air,[13] *its water having been poured out.*

DECLARATION

To remain emptied[13] is not to say vacuous[14] according to the tenth definition, but rather according to the eleventh; for in such case, the weight of the air would be negligible.

[11] *Nul corps.* Stevin's later use of the word *vuide*, though sometimes loose (*e.g.*, on pp. 149, n. 22, and 152) is not actually inconsistent with this definition, which gives it the meaning *vacuum*; but *vuidé* (Def. xi) does not mean evacuated. See Declaration, below.

[12] *Constitution,* in the sense of a mode of existence.

[13] *Vuidé.*

[14] *Vuide.*

POSTULATE VI

That the upper surface of the water (what is ordinarily called the fleur d'eau*) be plane and level, that is to say, parallel with the horizon.*

DECLARATION

It is known that every surface of water is spherical, or nearly so, and also that a part of it is a spherical part; but because if such surface should hereafter be taken to be spherical the demonstrations would be more difficult, and at the same time not practically useful; and because in practice we do not take the small sphere of water into consideration, we therefore count as negligible the contradiction that it would introduce into our postulate concerning the upper surface of water.

POSTULATE VII

If an upright column of water has its base and top parallel to the horizon and its lateral surface perpendicular thereto (so as to have all the straight lines between the corresponding points in the top and base perpendicular to the horizon) then it must be admitted that these perpendiculars tend towards the center of the earth, and that the base and the top may be parts of the surface of the earth.

DECLARATION

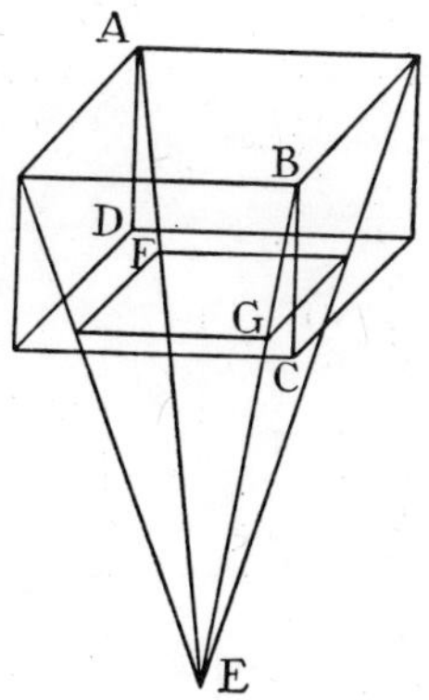

Let A B C D be water in the shape of an upright column, A B and C D parallel to the horizon, and A D and B C, etc., perpendicular thereto; also, let E be the center of the earth, from which the lines E F A, E G B, etc., are drawn, making F G similar to D C. Now, although it is true that A D and B C do not reach the center E, let it be granted that they lead to it, in view of the imperceptible difference that is found in practice, which is concerned only with lengths, surfaces, and bodies which have no perceptible share in the dimensions of the total earth. Also let it be granted that A B and C D are parts of the surface of the earth, namely, A B if the surface of the earth be [considered] thus far distant from the center E, and alternatively D C if the surface of the earth be sufficiently distant from the center.

The Propositions

THEOREM I—PROPOSITION I

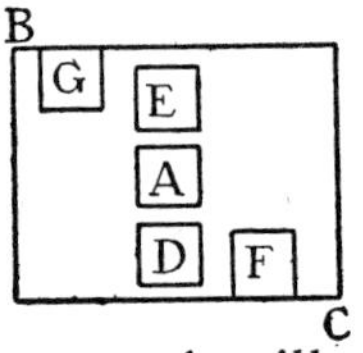

Any designated body of water maintains whatever position is desired in water.

Given. Let the water designated, in the *vasiforme* A, be put into the water B C.

Required. It must be demonstrated that the water A will remain there.

DEMONSTRATION

If it could be otherwise, namely, that A did not remain there but descended to the position D, then the water which took its place would descend lower for the same reason, and so of the remainder; so that this water would be in perpetual movement because of A; which is absurd.[15] And similarly, it may be demonstrated that A does not rise, nor move towards any side; but remains where it is put, whether in D, E, F, or G, or in any other place in the water B C.

Conclusion. Therefore the water designated, may be held in water wherever one desires; which was required to be proved.

THEOREM II—PROPOSITION II

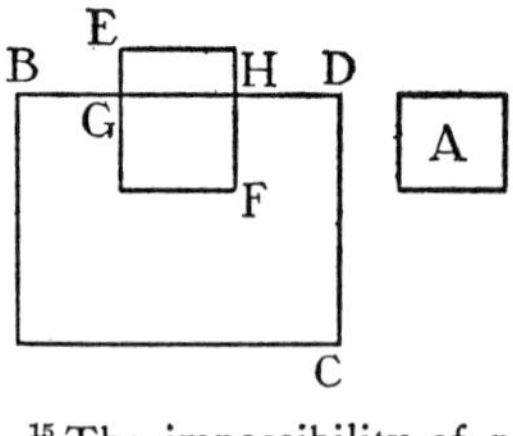

A solid body lighter than water does not sink altogether, but a part remains outside the water.

Given. Let the solid body A be lighter than the water B C, of which B D is the upper surface.

[15] The impossibility of perpetual motion in machines is the fundamental postulate of all Stevin's statics.

Required. It must be demonstrated that A, put into the water B C, will not be altogether submerged, but that there will be a part of it outside of the water.

Preparation. Let E F be a *vasiforme*, the part of which within the water and filled with water is G F, equal and similar to A; its surface G H will be in the surface B D, seeing that the *vasiforme* has neither weight nor lightness.

DEMONSTRATION

Since A is lighter than the water G F, by hypothesis, and G F is equal to A; G F then will be heavier than A: but if G F is emptied and A is put in its place, which is granted by the preparation, then, A being lighter than the water which was there, the *vasiforme* will not sink as before in the water, and therefore a part of A will be outside the water.

Conclusion. A solid body lighter than water, will not be altogether submerged in the water, but a part of it will be outside it; which was required to be proved.

THEOREM III—PROPOSITION III

A solid body multigrave *to water sinks to the bottom.*

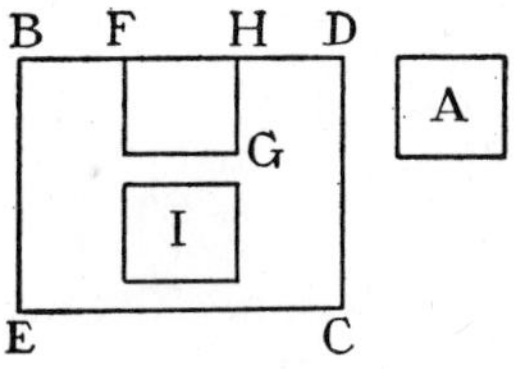

Given. Let A be a solid body *multigrave* to water; B D the water level, and E C the bottom.

Required. It must be demonstrated that A, in the water B C, will sink to the bottom, E C.

Preparation. Let F G be a *vasiforme*, full of water, equal and similar to A, the top of which, F H, is at the water level B D.

DEMONSTRATION

Because A is *multigrave* in comparison with the water F G, and F G is equal in size to A; A therefore will be heavier than F G. Let us empty the *vasiforme* F G, and in place of the water, let us put there the body A, which by hypothesis will exactly fill its place; then, since A is heavier than the water which was poured out, the *vasiforme* of A will sink further than that of the water F G, by the third postulate. We have now demonstrated that A will sink: and that it will sink to the bottom will be seen from what was said above; because it was demonstrated also that A could not hold itself in the position I, but that it would sink further than water in the same place. Therefore it will go to the bottom of the water.

Conclusion. A solid body *multigrave* to water will sink to the bottom; which was required to be proved.

THEOREM IV—PROPOSITION IV

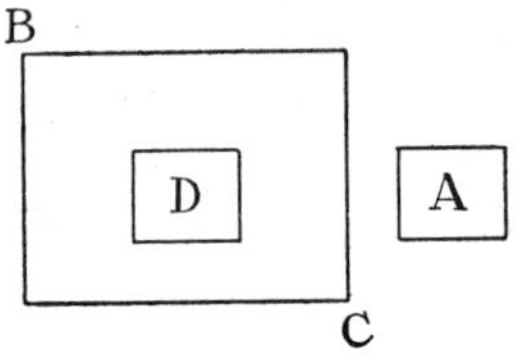

A solid body parigrave *to water holds itself in such position and place as is desired.*

Given. Let the solid body A be *parigrave* to the water B C.

Required. It must be demonstrated that A will hold itself in the water B C, in such place as is desired.

Preparation. Let D be a *vasiforme* full of water, equal to A.

DEMONSTRATION

Since A is *parigrave* with the water D, and D is equal in magnitude to A, then D will be *equiponderant* to A. Let us empty the water from the *vasiforme* D and let us put in its place A which is equal to it: the *vasiforme* D, then, will not sink further with A in it than it will with the water D in it, according to the third postulate. But the water D remains in any position in which it is put, by the Third Proposition. A, then, holds itself in the water B C in any desired place.

Conclusion. A solid body, then, *parigrave* with water, will hold itself in such position and place as is desired; which was required to be proved.

THEOREM V—PROPOSITION V

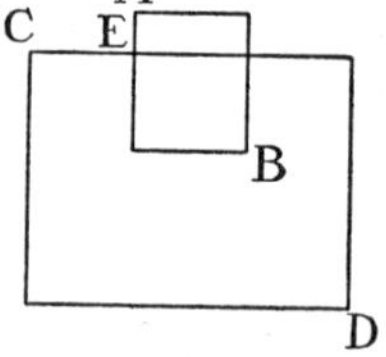

A solid body minugrave *to the water in which it lies is equiponderant to the water of which it takes the place.*[16]

Given. Let A B be *minugrave* to the water D C on which it floats, let its *vasiforme* be A B and its submerged part E B.

[16] Archimedes's Law of Floating Bodies. See reference on p. 19, n. 2.

Required. It must be demonstrated that the solid body A B is equiponderant to the water which is equal in volume to the part E B, which is in the water C D.

DEMONSTRATION

Let the solid body A B be removed from the *vasiforme* A B, and then let this *vasiforme* be refilled with water up to the place where the said *vasiforme* sinks in the water, just as before with the solid body. The water inside and outside is always of the same height, seeing that the *vasiforme* has neither weight nor lightness: hence, the water of the inside will weigh as much as the solid body A B, according to the third postulate.

Conclusion. A solid body *minugrave* to the water, etc.

PROBLEM I—PROPOSITION VI

A solid body of known size, floating on water of known density, having a part outside the water: To find the weight of the solid body.

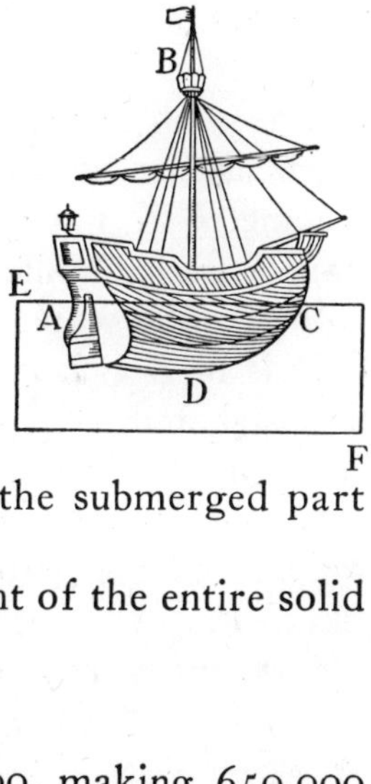

Given. Let A B C D be a solid body of any shape whatsoever; and let E F be water of which a cubic foot weighs 65 pounds (as much as a cubic foot of water weighs at Delft in Holland, determined by an experiment on which we base the following calculations); and let the submerged part of the solid body be 10,000 cubic feet.

Required. It is required to find the weight of the entire solid body A B C D and all that it supports.

OPERATION

We shall multiply 65 pounds by 10,000, making 650,000 pounds as required; the demonstration of which is shown by the preceding.

Conclusion. A solid body, therefore, of known size, floating on water of known weight, having a part outside the water: the weight of the whole solid body is determined as required.

THEOREM VI—PROPOSITION VII

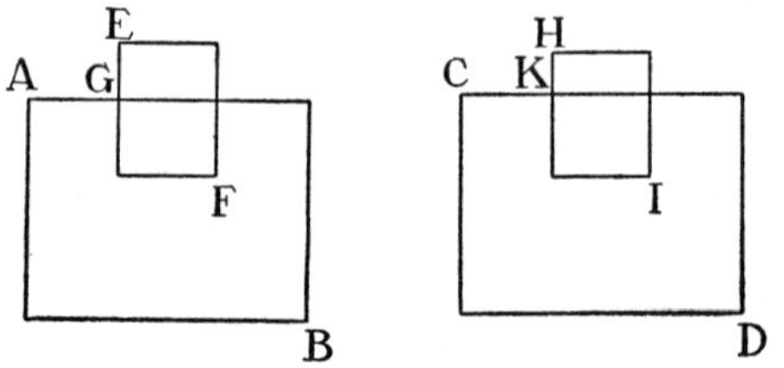

If a solid body is minugrave *to two* diversigrave *waters: as the weight of the* multigrave *water is to the other water, so will the volume of the solid body submerged in the* minugrave *water be to the volume submerged in the* multigrave.

Given. Let A B be water *multigrave* to the water C D, and E F a solid body *minugrave* to both of them; which put first in A B, having G F submerged, then in C D, having K I submerged.

Required. It must be demonstrated that as the weight of the *multigrave* water A B is to the weight of the *minugrave* water C D, so are the volumes K I and G F.

DEMONSTRATION

The *multigrave* water equal in volume to G F weighs as much as the body E F, and the *minugrave* water equal in volume to K I weighs as much as H I by the fifth proposition: but the bodies E F and H I are equal bodies by hypothesis; the water then of A B will be to G F as the water C D is to K I. But it results necessarily from the fifth definition, granted, that having two bodies G F and K I equiponderant their sizes will be proportional to their *gravities*, that is to say, to the densities of their matter: Therefore, as the weights of the waters A B and C D, so are the sizes K I and G F.

Conclusion. If a solid body, then, is *minugrave, etc.*

THEOREM VII—PROPOSITION VIII

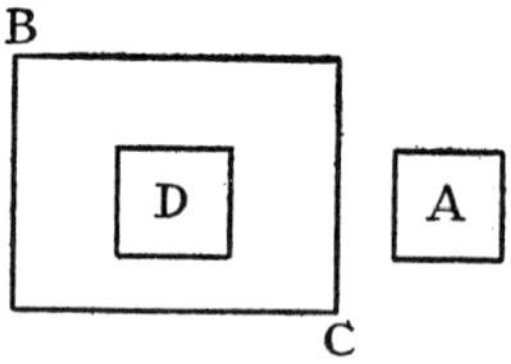

Every solid body is lighter in water than in air, by a weight of water equal in volume to itself.[17]

Given. Let A be a solid body, and B C the water.

[17] The Principle of Archimedes. See reference on p. 17, n. 1.

Required. It must be demonstrated that A, put into the water B C, will be lighter than it is in the air, by the weight of a body of water equal in volume to itself.

Preparation. Let D be a *vasiforme* similar and equal to A.

DEMONSTRATION

The *vasiforme* D full of water in the water B C is neither heavy nor light seeing that it remains wherever it is put, according to the first proposition: wherefore, the water in D, having been removed when the body A is put in its place, it will be found to be of the lightness mentioned, namely, the weight of A less the weight of the water emptied out. But such water is equal to A in volume: A, then, in the water B C, will be found to be lighter than it is in the air by the weight of a volume of water equal to itself: which was required to be proved.

PROBLEM II—PROPOSITION IX

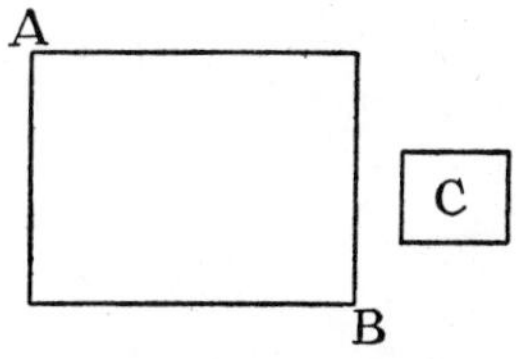

Being given the quantigravité[18] *of water and of a solid body, and also the weight of the same body: To find its condition*[19] *in the water.*

First case: that of a solid body *minugrave* to water.

Given. Let A B be the water, and C a solid body weighing two pounds, and let the *quantigravité* of water and that of the solid body be as 5 to 1.

Required. It is required to find the condition of C in the water.

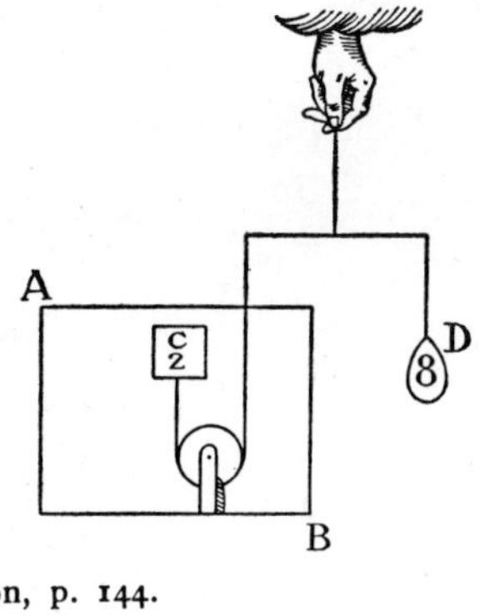

OPERATION

One first determines the weight of a volume of water equal to C; this will be found to weigh five times 2 pounds, or 10 pounds. Subtracting from this weight 2 pounds (the weight of the solid body), there will

[18] Density. See above, Theorem VI, Demonstration, p. 144.

[19] *Constitution.* See p. 138, n. 12.

remain 8 pounds' diminution of weight, that is to say, of lightness or buoyancy, for C in the water. This could be shown more plainly by putting C in the water A B, and balancing it by a weight of 8 pounds, D, as in the accompanying illustration, when they will be in equilibrium.

Second case: that of a solid body multigrave *to water, when the operation is as before.*

Given. Let the *quantigravité* of the water A B be to that of the solid body C as 1 to 4: and let C weigh 12 pounds.

Required. It is required to find its condition in the water A B.

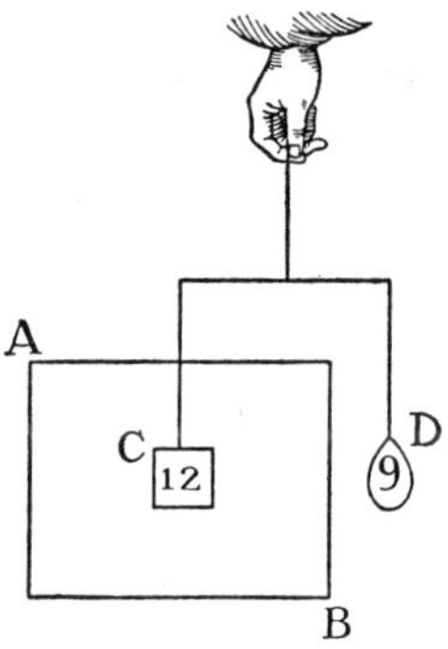

OPERATION

One first determines how much a body of water equal in volume to C will weigh; and it will be found to be one-quarter the weight of C, 12 pounds, which is 3 pounds. This, taken from 12 pounds, the weight of C, leaves 9 pounds for the weight of C in the water.

This case also we could understand, by placing the 9-pound weight so that it balances the 12 pounds of C in the water, as in the adjoining illustration.

We could cite, further, a third case in which the water and the solid body are *parigrave*, but it is plain that a body such as this would be neither lighter nor heavier in the water, by the eighth proposition.

Conclusion. Being given the *quantigravité* of water and of a solid body, also the weight of the said body, we have from this found its condition in water: which was to be done.

THEOREM VIII—PROPOSITION X

The bottom of a mass of the water, parallel to the horizon, supports a weight, equal to the weight of a column of water of which the base is the aforesaid bottom and the height a line perpendicular to the horizon, between the bottom and the surface of the water.

Given. Let A B C D be water, in the shape of a rectangular parallelopiped, A B its upper surface, E F the level bottom, and E G the perpendicular between the bottom E F and the surface of the water. Let the column be that which is held between the bottom for its base E F, and its height G E, namely, the column G E F H.

Required. It must be demonstrated that the bottom E F supports a weight equal to the weight of the column of water G E F H.

DEMONSTRATION

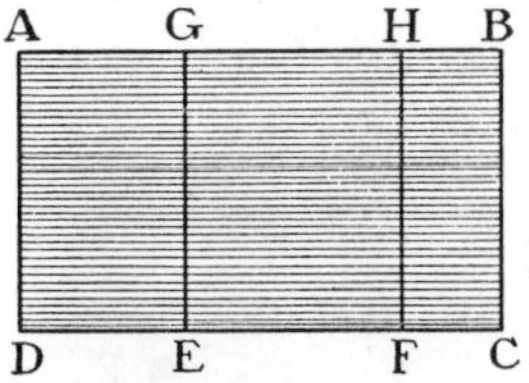

If on the base E F there rests a weight greater than G E F H, it would be due to the neighboring water. Let it be, if it were possible, due to the water A D E G, and H F C B; then, in the same way it could be said that there rests on the bottom D E more than the water A D E G, and on F C more than the water H F C B; so that on D C there would rest more than the water A D C B; which is absurd, that being a rectangular parallelopiped. Similarly it may be demonstrated that on E F the weight will not be less than G E F H; so that, consequently, on E F there will rest a weight precisely equal to the column of water G E F H.

COROLLARY I

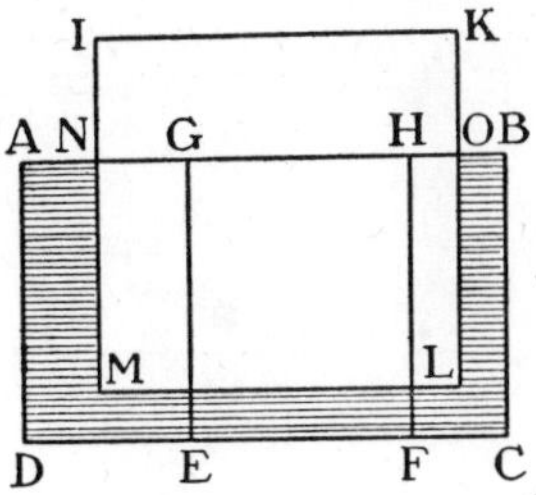

In the water A B C D of the tenth proposition, let us put I L, a solid body *minugrave* to the water, that is to say, floating on the water—the part N L being in the water, and N K outside it, as in the adjacent illustration. Then the solid I L will be equiponderant to the water N O L M, by the fifth proposition, from which it follows that the body I L, with the rest of the water around it, is equiponderant to a body of water equal to A B C D; wherefore, let us say again, according to the proposition, there rests on the bottom E F a weight equal to the weight of a column of water such that G E F H, of which E F is the base, and G E a perpendicular between the bottom and the surface of the water is the height. From this we may conclude that when a substance floats on the water, it makes no change in the weight sustained by the bottom when the water remains at the same height.

COROLLARY II

Again, let there be in the water A B C D one or several solid bodies *parigrave* to the water, so that there is room only for the water I K F E L M. Then this body (or bodies) neither increases nor decreases the weight resting on the base

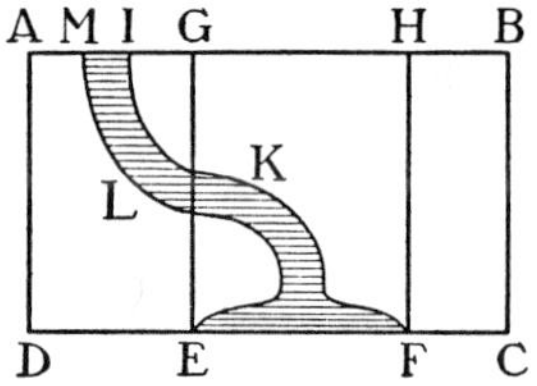

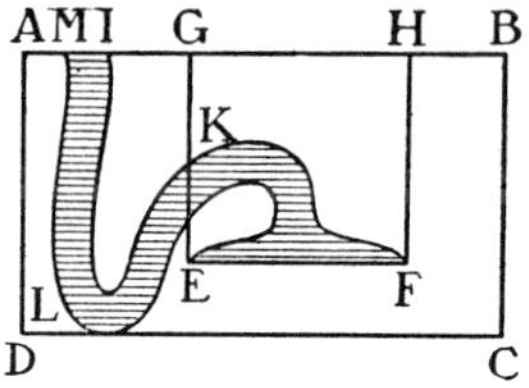

E F. And therefore, according to the proposition, on the bottom there rests a weight equal to the weight of the water in a column having the said bottom for base, and a height equal to that perpendicular to the horizon, which lies between the bottom and the surface of the water.

COROLLARY III

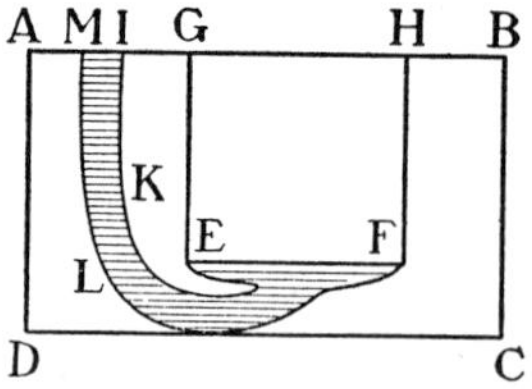

Again, let A B C D be entirely water, and E F a base within it parallel to the horizon, then the water will press[20] as much from below in rising as from above in falling. Otherwise the stronger pressure would prevail over the weaker; but this does not happen, for in accordance

[20] *Poussera.*

with the first proposition everything maintains the disposition that was given to it. Now let there be in the water some solid bodies *parigrave* to it, and so disposed that the water I K E F L M presses E F from below—namely, against the solid body, as previously against the water, both pressing against each other equally; then, against E F there is a force[21] which presses it upward, exactly as the column of water G E F H presses the same base downward, according to the proposition; for the height G E is the perpendicular between the surface of the water and the bottom E F.

COROLLARY IV

If the solid bodies of the second and third corollaries were put in their places, and the water emptied, leaving in its place a vacuum[22] I K F E L M; then the bottom E F would bear no weight. And if one refilled the emptied place with water, the bottom would sustain as much force as though the vessel were entirely filled with water (having lost the solid bodies).

COROLLARY V

But if the unnecessary substances, the solid bodies, are taken away, and if there is no longer anything to hold the

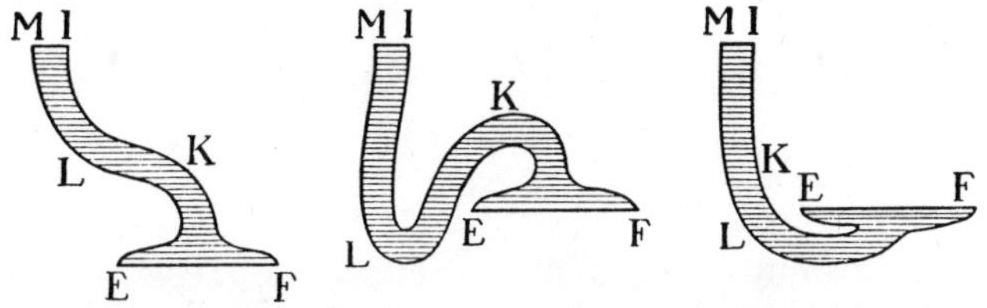

water in shape, then the bottom E F will sustain the same force as if there were a column of water above it, having the same bottom as base, of which the height would be the perpendicular between the said bottom and the surface of the water.

[21] *Effort.*
[22] *Vide.*

Conclusion. On the bottom of the water, parallel with the horizon, there rests a weight, etc.

Read the proofs deduced in the Appendix on the Practice of Hydrostatics.

NOTE

We could also enunciate the tenth proposition as follows: *At the bottom of any body of water, on a surface parallel to its upper surface, rests a mass equal in weight to the water contained in a spherical sector truncated by a spherical surface parallel to or homocentric with the spherical surface of the earth.*

We might have made the corresponding demonstrations, also, as above; but have not done so for the reasons stated in the seventh postulate.[23]

FIFTH BOOK OF STATICS

Commencing the Practice of Hydrostatics[1]

The weight of a boat, including all of it which is above the water, or that of any body floating on the water, being made sufficiently evident by the Sixth Proposition,[2] we shall ignore this subject and shall consider the consequences of the Seventh Proposition.

[23] Translator's note: The rest of this fourth book of the Statics consists of interesting geometrical demonstrations whereby Stevin determines the centers of pressure on surfaces variously disposed obliquely to the horizon: its content, not directly relevant to the present discussion, is here omitted. That of the fifth book, on the other hand, exhibits anticipations of Pascal's generalizations, and on this account the whole of it is given below.

[1] A note "To the Reader" is here omitted.

[2] Page 143.

PROPOSITION I

To find how much deeper a given body minugrave *to water will sink in one sort of water than in another more dense.*[3]

Suppose, for example, that we wish to know how much more a boat will sink in the water before Catwijck, than in the River Rhine at Leyden. For this purpose we shall examine the *quantigravité* of the waters, which, let us say, are, the fresh to the salt, as 42 to 43. This I have found them actually to be in the month of July; because, by taking two equal bodies,[4] the water of the Rhine was found to weigh 4,260 grains, but that of the sea 4,362 grains, which is very nearly as 42 is to 43.

Therefore one may say that the volume of the submerged part of a boat in the Rhine will be to the submerged part in the sea as 43 to 42. From which, the geometrician will be able to judge how much further it[5] will sink in the one water than in the other. The necessity of the rule here used depends upon the Seventh Proposition of the Elements of Hydrostatics.[6]

PROPOSITION II

To disclose what is really contained in the Tenth Proposition of the Hydrostatics.

We have shown by the Tenth Proposition aforesaid, in the Fifth Corollary, that the base beneath the water, there designated E F, is not more burdened by much water than it is by little (the height remaining the same), but equally. And inasmuch as many would consider this to be contrary to nature, we shall here set forth (what we have already demonstrated mathematically) an example in practice, by which it may be understood and better comprehended.

EXAMPLE 1

Let the base A B be equal to C D, let the heights E F and G H be equal, and let the water in E I F be less than that in G C D, in such ratio that E A B may be 1 pound and G C D 10 pounds.

[3] *Pesante.*

[4] *Corps egaux;* here, equal volumes of these waters.

[5] That is, the boat, or other body *minugrave* to water.

[6] Page 144.

Suppose that the figure G C D represents a round column, so that G C D will contain 10 times more water and will be as many times heavier than E A B; this notwithstanding, the water presses the bottom A B as much as the water G C D presses the bottom C D. This may be demonstrated by experiment in the following manner.

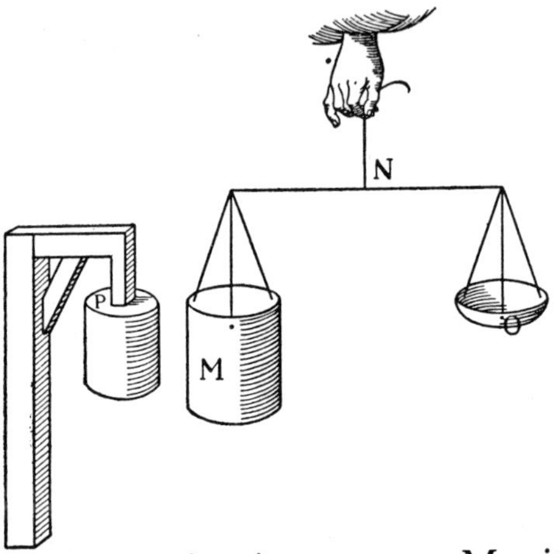

Let M N O be a balance, the pans being M and O, of which let M be in the form of a column similar to the vessel G C D named above, and containing 10 pounds of water. At the same time, let P be a body of wood, solid and fixed, as in the adjoining illustration, and similar to M, but smaller so that it can enter M without touching it.

Now let the pan M be put around P, as in the second figure following, and in the pan O, let a weight of 10 pounds, Q, be put; then M will press the lower part of P as much as the weight Q is heavy. Suppose now that the space (*vuide*) between P and M can be filled with a pound of water, i.e., by a body of water equal in weight to the body E A B. Then the pound of water in the space will cause the pan M to descend and O to rise, as the experiment will show, and also as the reasoning sets it forth in the Tenth Proposition referred to. In this way, therefore, 1 pound of water

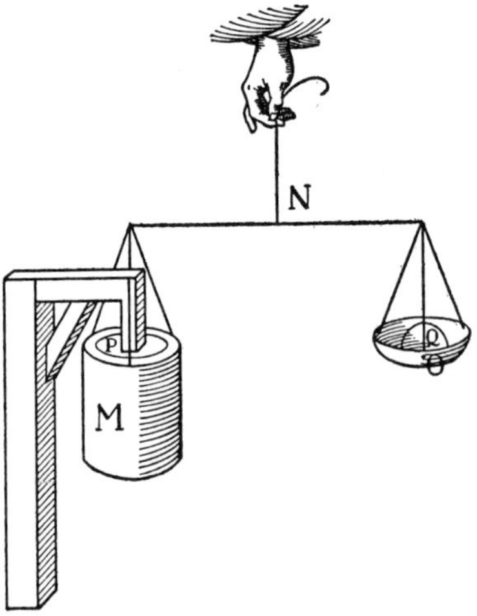

in M will have as much force as 10 pounds in the pan M, let it be of iron, of brass or copper, or of whatever matter it may be. And for the same reason, 1 pound of water may exert as much force as 1,000 pounds of other weight. This being so, there will be water between the pan M and the under part of P against which the bottom of M presses as much as it formerly pressed against the same part before the water was put into it, namely, with the weight of Q, this being 10 pounds. Thus the bottom of M presses as much against the water as the 10 pounds of Q can press; and, on the contrary, the water presses as much against the bottom of M as the 10 pounds of Q can press. Let us now assume that the water resting against the bottom M, is equal to the water K L B A, and that the remainder which is around P is equal to the water I E; then the water E A B will press as much against the bottom A B, as it presses against the bottom M; namely, with a pressure equal to that of 10 pounds; but this is the pressure of the water in G C D against the bottom C D. It follows then that the water of the vessels E A B and G C D press equally against their bases A B and C D, that is to say, 1 pound presses as much as 10 pounds. And in the same way, that 1 pound may be made to press as much as a thousand pounds.

EXAMPLE 2

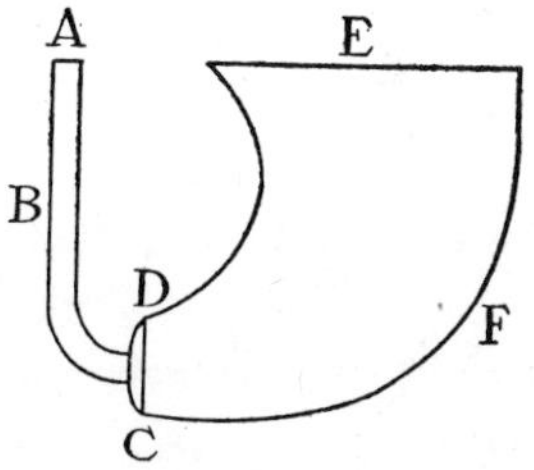

Let A B C D be a slender tube, and C D E F a large vessel separated from the tube by the bottom C D, both filled with water so that the surfaces of the two waters may be in the same plane. Now, if the base were removed and the water C E were to press more than the water A D C, then the weaker would give place to the stronger and consequently the water A D C would move away from this plane, which would contradict experience; from which it is concluded that the small body of water A D C presses as much against the base D C as the larger body of water C E.

EXAMPLE 3

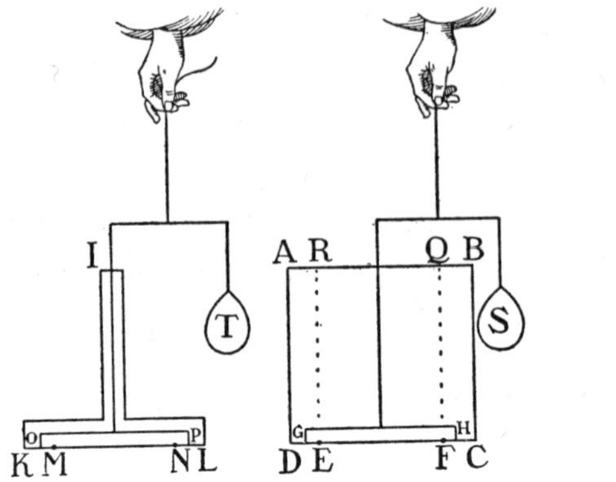

Let A B C D be a vessel full of water, at the bottom of which (this being level) there is a round opening E F, which is covered by a plate of wood *minu-grave* to the water. Let also I K L be another vessel of the same height, and with an equal opening in the bottom, and provided also with a plate of wood similar and of equal weight with the other, which nicely closes the opening. It will be found by test that the plates will be equally pressed upon, and will not rise, as wood would ordinarily rise in water, but will press equally on the bottoms; which may be recognized experimentally by attaching equal, suspended weights, T and S, equiponderant to the water supported by the plate G H, that is, to the column of water E R Q F.

NOTE

It is evident that if the difference between the weight of the plate, such as G H, and that of the water of the same volume would support the weight of the column of water E F Q R, such a plate would not rest against the opening E F, but would be raised to the top of the water.

It is also evident that if such a plate were also *multigrave* to water, like lead, iron, etc., it would press the bottom as much as the weight of the column of water E F Q R, and more also, because the plate would weigh more than an equal volume of water.

But if the plate were *parigrave* to water, it is certain that it would press as much against the opening E F, as would a weight equal to that of the column of water E F Q R.

EXAMPLE 4

Let A B C D be a vessel full of water, with an opening E F in the bottom C D, on which there rests a plate *minugrave* to the water; this plate will press the bottom as has been said above. Then let there be at I K L a little tube, of which the upper opening I is of the same height as A B; and let its lower opening be E F. Now, when this tube is filled with water, the small volume of this water will press as much against the plate from beneath as the greater volume presses it from above; because the plate G H will then be raised to the top. So that 1 pound of water (granting that the tube I K L contains that amount) will exert more pressure against the plate G H, than 100,000 pounds would, such as S above [in Example 3]; something that would be considered a mystery in nature if the cause were unknown.

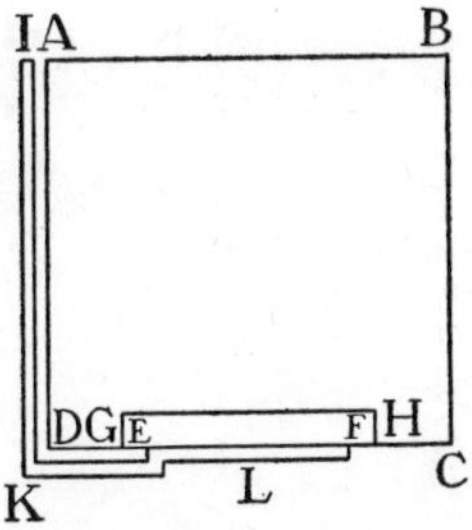

EXAMPLE 5

Now, to give a very clear illustration of cases in which the water above presses against the bottom, like that discussed in the Third Corollary of the Tenth Proposition mentioned: Let A B C D be a body of water, and E F a tube, G a plate *multigrave* to water, like lead, etc., as in the first figure here shown.

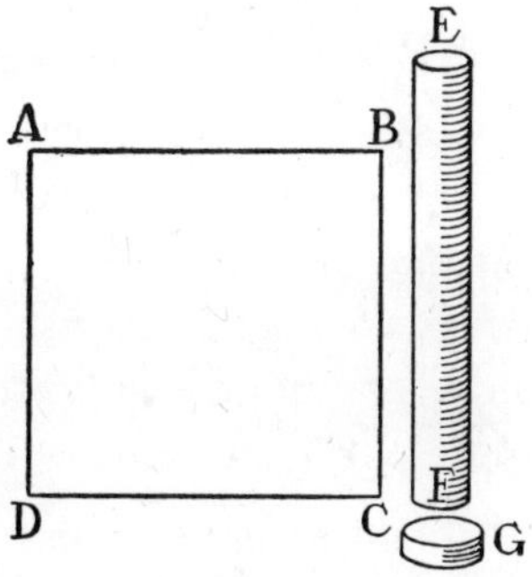

Suppose the plate G to be placed against the opening F, closing it exactly, and the tube thus joined to the plate to be put in the water A B C D, say as far as H, as is shown in the adjoining illustration. The plate G will not fall to the bottom as lead would usually fall, but will

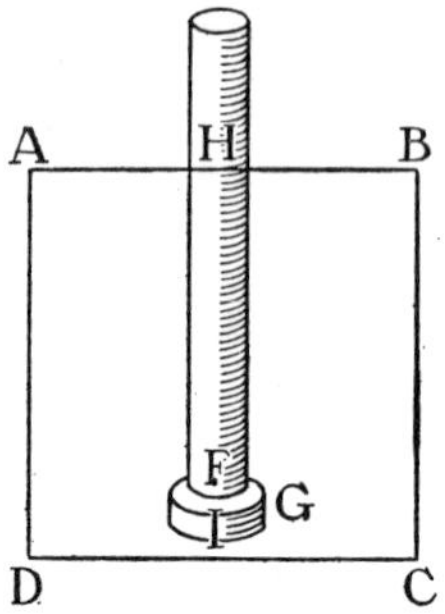

remain suspended on the tube, pressing against it, as would a column of water having the base equal to the opening F, and the height H I, less the height determined by the difference between [the weight of] the plate G and [that of] an equal body of water: but if the plate were not well fitted to the opening, the water would enter there, and the plate would remain suspended until the water which entered had gained the said weight.

But if anyone should think that the plate would be held longer suspended on the tube when this is in a greater volume of water, he would find that in reality it would be the same as in a small quantity of water, as illustrated.

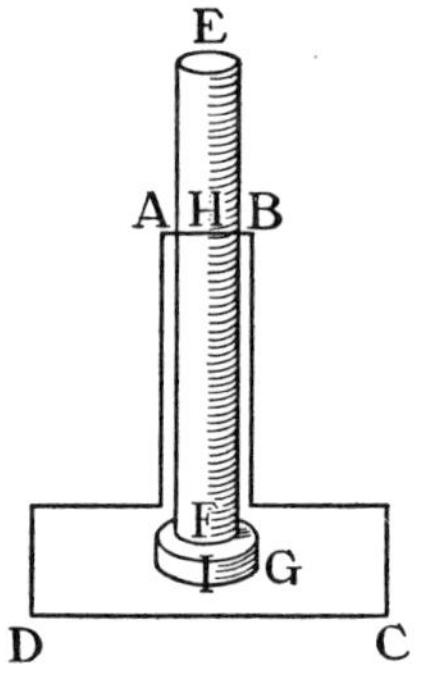

Conclusion. We have then set forth the import of the Tenth Proposition, as designed.[7]

NOTE

Regarding the Eleventh Proposition,[8] by which, among other things, the force of water pressure against the gates of dams is explained, and it is shown that the water on one side of them, even though its area is very small, will press as much against them as the great Ocean on the other side, provided the waters be of the same height—all this, being clearly evident, will be omitted.

[7] It is clear that this discussion of the implications of proposition X explains completely the "hydrostatic paradox" and comes as near to demonstrating the action of the hydrostatic press as would be possible without building the actual machine. The injustice of the common ascription of these discoveries to Pascal is therefore obvious.

[8] Omitted in this excerpt.—Translator.

PROPOSITION III

To show the reason why a man, swimming at very deep levels, does not die from the effects of the great quantity of water above him.

Let a man be at a depth of 20 feet in the water, the foot of water weighing 65 pounds, and the entire surface of his body being 10 feet. This being so, 13,000 pounds will press against his body, by the Tenth and Eleventh Propositions of the Elements of [Hydrostatics, in the Fourth Book of] Statics:[9] therefore, it may be asked how it is possible that a person does not burst from so great a load? To which the response will be as follows:

A.[10] *All pressure which crushes the body, presses some portion of the body out of its natural place.*

O. *The pressure caused by water presses no part of the body out of its natural place.*

O. *The pressure caused by the water, therefore, does not crush or hurt the body.*

The minor [premise] is shown by experience, the reason being that if there were something which was being pressed out of its place, it would necessarily enter another place. Now, this place is not outside, because the water presses on all sides equally (though the part below is a little more pressed than that above, by the Eleventh Proposition of the Elements of Hydrostatics. This need not be considered, since such difference can not press any part of the body out of its natural place). But neither is this place inside the body, because there is no empty space there, any more than there is outside; from which it follows that the particles press against one another equally, since the water has the same natural action all around the body. This place, then, is neither inside nor outside the body, and consequently is nowhere; the result being that no part is pressed out of its natural place and therefore that the body is not hurt at all.

[9] Pages 146 ff.

[10] These letters refer to the particular figure of the syllogism.

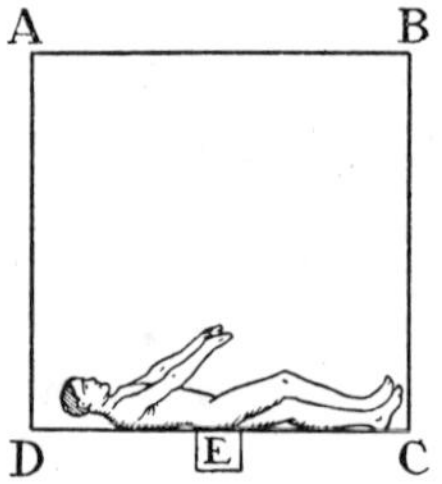

To set this forth more clearly, let A B C D be a body of water, having in the base D C [of its container] an opening closed by a pin. On this base lies a man F, having his back on E. This being the situation, since the water is pressing him on all sides, that which is above him presses no part of his body out of its place.

But if one wishes to see the real cause of this effect, it would only be necessary to take away the pin E. Then there would be no pressure against his back at E, similar to the pressure against the other parts of his body, and consequently his body would suffer there a compression, the strength of which was demonstrated in the Third Example of the Second Proposition of the present book: namely, a pressure equal to that of a column of water having the opening E for base, and A D for height; which demonstrates clearly that which was intended.

END OF THE FIFTH BOOK

Appendix II

GALILEO'S REMARKS ON NATURE'S ABHORRENCE OF A VACUUM[1]

[In this Dialogue of the First Day, Galileo discusses the strength of materials; and in this connection considers nature's abhorrence of the vacuum. He finds considerable difficulty in assigning to this conception as such any cogent meaning, but accepts it tentatively as referring to a well-known type of natural effect, and seeks to give it a definite physical significance by measurement. To this end he devises an experiment whereby he determines the breaking strength of a short column of water, which fills a tube closed at the top and is held in place by a tightly fitting piston below, which may be weighted at pleasure. Since the cohesion thus seemingly demonstrated cannot be ascribed to any physical property of the mobile substance, he accepts it as measuring a real effect which he identifies with that commonly known as the *horror vacui*; and then shows, by reference to the behavior of pumps, that this effect is, within his experience, invariable.

The following excerpts from Galileo's discourse are sufficient to indicate the character of his reasoning upon this matter. It will be noted that the animistic implications which Pascal twenty years later still associates with the idea of *horror vacui* and takes the trouble emphatically to repudiate, are already ignored by Galileo, apparently as not worth discussion; his argument showing that this *horror vacui* is merely a metaphorical expression for a real and determinable physical effect.]

[1] Quoted from Crew and De Salvio's translation of the *Discorsi e Dimostrazioni Matematiche intorno a Duo Nuove Scienze* (1638), *Giornata Prima,* published under the title, *Dialogues concerning Two New Sciences, by Galileo Galilei,* New York, The Macmillan Company, 1914, pp. 11, 12-13, 16-17.

SALVIATI: . . . First I shall speak of the vacuum, demonstrating by definite experiment the quality and quantity of its force (*virtu*). If you take two highly polished and smooth plates of marble, metal, or glass and place them face to face, one will slide over the other with the greatest ease, showing conclusively that there is nothing of a viscous nature between them. But when you attempt to separate them and keep them at a constant distance apart, you find the plates exhibit such a repugnance to separation that the upper one will carry the lower one with it and keep it lifted indefinitely, even when the latter is big and heavy.

This experiment shows the aversion of nature for empty space, even during the brief moment required for the outside air to rush in and fill up the region between the two plates. . . .

SAGREDO: Allow me to interrupt you for a moment, please; for I want to speak of something which just occurs to me. . . . The fact that the lower plate follows the upper one allows us to infer, not only that motion in a vacuum is not instantaneous, but also that, between the two plates, a vacuum really exists, at least for a very short time, sufficient to allow the surrounding medium to rush in and fill the vacuum; for if there were no vacuum, there would be no need of any motion in the medium. One must admit, then, that a vacuum is sometimes produced by violent motion (*violenza*) or contrary to the laws of nature (although in my opinion nothing occurs contrary to nature except the impossible, and that never occurs).

But here another difficulty arises. . . . For the separation of the plates precedes the formation of the vacuum which is produced as a consequence of this separation; and since it appears to me that, in the order of nature, the cause must precede the effect, even though it appears to follow in point of time, and since every positive effect must have a positive cause, I do not see how the adhesion of two plates and their resistance to separation—actual facts—can be referred to a vacuum as cause when this vacuum is yet to follow. According to the infallible maxim of the Philosopher, the non-existent can produce no effect.

SIMPLICIO: Seeing that you accept this maxim of Aristotle, I hardly think you will reject another excellent and reliable maxim of his, namely, Nature undertakes only that which happens without resistance; and in this saying, it appears to me, you will find the solution of your difficulty. Since nature abhors a vacuum, she prevents that from which a vacuum would follow as a necessary consequence. Thus it happens that nature prevents the separation of the two plates.

. . . .

SAGREDO: Thanks to this discussion, I have learned the cause of a certain effect which I have long wondered at and despaired of understanding. I once saw a cistern which had been provided with a pump under the mistaken impression that the water might thus be drawn with less effort or in greater quantity than by means of the ordinary bucket. The stock of the pump carried its sucker and valve in the upper part so that the water was lifted by attraction and not by a push as is the case with pumps in which the sucker is placed lower down. This pump worked perfectly so long as the water in the cistern stood above a certain level; but below this level the pump failed to work. When I first noticed this phenomenon I thought the machine was out of order; but the workman whom I called in to repair it told me the defect was not in the pump but in the water, which had fallen too low to be raised through such a height. And he added that it was not possible, either by a pump or by any other machine working on the principle of attraction, to lift water a hair's breadth above eighteen cubits; whether the pump be large or small this is the extreme limit of the lift. Up to this time I had been so thoughtless that, although I knew a rope, or rod of wood or of iron, if sufficiently long, would break by its own weight when held by the upper end, it never occurred to me that the same thing would happen, only much more easily, to a column of water. And really is not that thing which is attracted in the pump a column of water attached at the upper end stretched more and more until finally a point is reached where it breaks, like a rope, on account of its excessive weight?

SALVIATI: That is precisely the way it works; this fixed elevation of eighteen cubits is true for any quantity of water whatever, be the pump large or small or even as fine as a straw. We may therefore say that, on weighing the water contained in a tube eighteen cubits long, no matter what the diameter, we shall obtain the value of the resistance of the vacuum in a cylinder of any solid material having a bore of this same diameter.

APPENDIX III

TORRICELLI'S LETTERS ON THE PRESSURE OF THE ATMOSPHERE

Translated from the text of Loria and Vassura[1] by Vincenzo Cioffari.

---

1. *Letter of Torricelli to Michelangelo Ricci*

Florence, June 11, 1644.

MY MOST illustrious Sir and most cherished Master: Several weeks ago I sent some demonstrations of mine on the area of the cycloid to Signor Antonio Nardi, entreating him to send them directly to you or to Signor Magiotti after he had seen them. I have already intimated to you that a certain physical experiment was being performed on the vacuum; not simply to produce a vacuum, but to make an instrument which would show the changes in the air, which is at times heavier and thicker and at times lighter and more rarefied. Many have said that a vacuum cannot be produced,

[1] Gino Loria e Giuseppe Vassura: *Opere di Evangelista Torricelli* (Faenza, 1919), III, 186 *seq.*

others that it can be produced, but with repugnance on the part of Nature and with difficulty; so far, I know of no one who has said that it can be produced without effort and without resistance on the part of Nature. I reasoned in this way: if I were to find a plainly apparent cause for the resistance which is felt when one needs to produce a vacuum, it seems to me that it would be vain to try to attribute that action, which patently derives from some other cause, to the vacuum; indeed, I find that by making certain very easy calculations, the cause I have proposed (which is the weight of the air) should in itself have a greater effect than it does in the attempt to produce a vacuum. I say this because some Philosopher, seeing that he could not avoid the admission that the weight of the air causes the resistance which is felt in producing a vacuum, did not say that he admitted the effect of the weight of the air, but persisted in asserting that Nature also contributes at least to the abhorrence of a vacuum. We live submerged at the bottom of an ocean of the element air, which by unquestioned experiments is known to have weight,[2] and so much, indeed, that near the surface of the earth where it is most dense, it weighs [volume for volume] about the four-hundredth part of the weight of water. Those who have written about twilight, moreover, have observed that the vaporous and visible air rises above us about fifty or fifty-four miles; I do not, however, believe its height is as great as this, since if it were, I could show that the vacuum would have to offer much greater resistance than it does—even though there is in their favor the argument that the weight referred to by Galileo applies to the air in very low places where men and animals live, whereas that on the tops of high mountains begins to be distinctly rare and of much less weight than the four-hundredth part of the weight of water.

We have made many glass vessels like the following marked A and B with necks two cubits.[3] We filled these with quicksilver,

[2] See note, p. 73, n. 5.

[3] *Braccia;* each closely 23 inches. The "vessels" A and B, according to Torricelli, are the parts *above* the "necks" BC, AD, in the figure (p. 165).

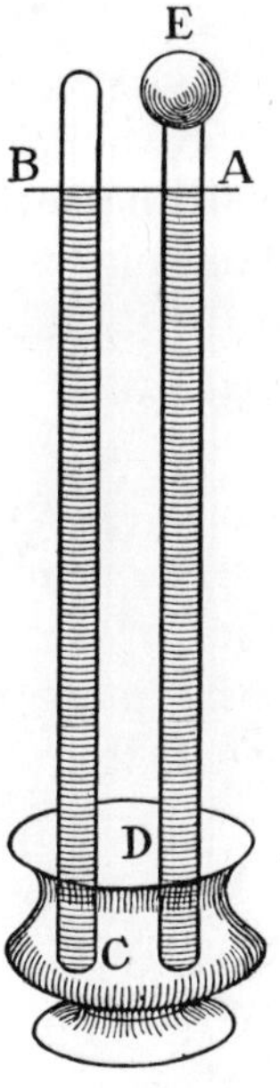

and then, the mouths being stopped with a finger and being inverted in a basin where there was quicksilver C, they seemed to become empty and nothing happened in the vessel that was emptied; the neck AD, therefore, remained always filled to the height of a cubit and a quarter and an inch besides. To show that the vessel was perfectly empty, the underlying basin was filled with water up to D, and as the vessel was slowly raised, when its mouth reached the water, one could see the quicksilver fall from the neck, whereupon with a violent impetus the vessel was filled with water completely to the mark E. This experiment was performed when the vessel AE was empty and the quicksilver, although very heavy, was held up in the neck AD. The force which holds up that quicksilver against its nature to fall down again, has been believed hitherto to be inside of the vessel AE, and to be due either to vacuum or to that material [mercury] highly rarefied; but I maintain that it is external and that the force comes from without. On the surface of the liquid which is in the basin, there gravitates a mass of air fifty miles high; is it therefore to be wondered at if in the glass CE, where the mercury is not attracted nor indeed repelled, since there is nothing there, it enters and rises to such an extent as to come to equilibrium with the weight of this outside air which presses upon it? Water also, in a similar but much longer vessel, will rise up to almost eighteen cubits, that is, as much further than the quicksilver rises as quicksilver is heavier than water, in order to come to equilibrium with the same force, which presses alike the one and the other.

The above conclusion was confirmed by an experiment made at the same time with a vessel A and a tube B, in which the quicksilver always came to rest at the same level, AB. This is an almost certain indication that the force was not within;

because if that were so, the vessel AE would have had greater force, since within it there was more rarefied material to attract the quicksilver, and a material much more powerful than that in the very small space B, on account of its greater rarefaction. I have since tried to consider from this point of view all the kinds of repulsions which are felt in the various effects attributed to vacuum, and thus far I have not encountered anything which does not go [to confirm my opinion]. I know that you will think up many objections, but I also hope that, as you think about them, you will overcome them. I must add that my principal intention—which was to determine with the instrument EC when the air was thicker and heavier and when it was more rarefied and light—has not been fulfilled; for the level AB changes from another cause (which I never would have believed), namely, on account of heat and cold; and changes very appreciably, exactly as if the vase AE were full of air.[4]

2. *Extract from Ricci's letter to Torricelli in response to the preceding*[1]

Rome, June 18, 1644.

. . . First of all, it seems to me that one could exclude the action of the air in gravitating on the outer surface of the quicksilver in the basin by placing on this vessel a cover pierced with a single hole through which the glass tube passes and then stopping all parts completely so that there will be no further communication. The air above the basin would, in such a case, gravitate no longer on the surface of the quicksilver, but upon the cover; and if then the quicksilver remains suspended in the air as before, the effect could no longer be attributed to the weight of the air which is supposed to hold it there in a sort of equilibrium. Secondly, if we take a syringe (which should be frequently used in this sort of inquiry and should have its sucker completely enclosed so as to exclude with its bulk every other body), and if then we stop the hole on top and pull the piston back by force, we feel a great resistance; and that effect follows not only when the syringe is held downwards so that the sucker

[4] Thus acting like Galileo's thermometer.

[1] Loria e Vassura, *op. cit.*, III, 194, l. 3 *seq.*

is brought above and the air gravitates upon the top of the piston rod, but it follows in whatever direction the syringe is turned. In these cases, it is still not evident that one can easily imagine how the weight of the air has anything to do with the effect. Finally, a body submerged in water is not counterpoised[2] by all of the water above it, but by that alone which is displaced when the body is immersed, [the mass of] which is not greater than [the mass of] that body; and since I should think that the same principle would apply to the balancing of the quicksilver, should this not be counterbalanced[3] by as much air as equals its own mass, and if so how could the air ever preponderate? . . .

3. *Torricelli to Ricci*

Florence, June 28, 1644.

I am not unaware of the high prerogatives and great merits of the most illustrious Cavalier Del Pozzo, a real possessor and protector of virtue and of the virtuous. Would that I, as a simple lover of virtue, could have enjoyed the benefits of his powerful protection at the opportune time, for then I might reasonably have hoped for other advancements than such as I could look for from poor D. Benedetto, may he rest in peace. As for supplying you with lenses, it is not possible to do so, because I have none; and I cannot possibly complete the work before next October, when I shall have finished my most tedious printing of the little book, which torments me ten times more than its first conception and composition did. Moreover, during the heat one cannot stand so much work, and I am doing everything by myself, since I want no help or service in this business. Some people from the other side of the mountains, passing through Florence, have begged me [for lenses] and offered me many pistoles for them. God only knows how willingly I would have taken this money; yet I could not be of service to them. I worked a few weeks, that is during the whole of Lent and fifteen days of Carneval, and made four or five perfect lenses. The Grand Duke has some of them, the Princes others, and some I have given to

[2] *Contrasta,* not necessarily implying equilibrium.

[3] *Contrastare,* though here there is an equilibrium.

other Patrons. I myself have kept two of them, since they are not, as I would wish, of the sort that can meet competition: if I survive and have the time and the material, I shall keep your request in mind.

Meanwhile I consider it superfluous to answer your three objections to my theory on the resistance met in producing the vacuum, because I expect that the explanations will have come to you yourself since you wrote the letter. As for the first, I answer: If you introduce a soldered sheet of metal to cover the surface of the basin, placing it in such a way that it touches the quicksilver in the basin, that quicksilver which is raised in the neck of the vessel will remain, as before, suspended; not by the weight of the sphere of the air, but because the part in the basin cannot give way. If, on the other hand, you introduce this sheet so that it will enclose some air also, I would ask whether or not you wish that contained air to be of the same degree of density as the outer air. If so, the quicksilver will be held up as before (after the manner of the wool which I shall mention presently); but if the air which you include is to be more rarefied than the outer air, then the raised metal will descend somewhat; moreover, if it were infinitely rarefied—that is, a vacuum—then the metal would descend entirely, granted that the enclosed space could contain it.

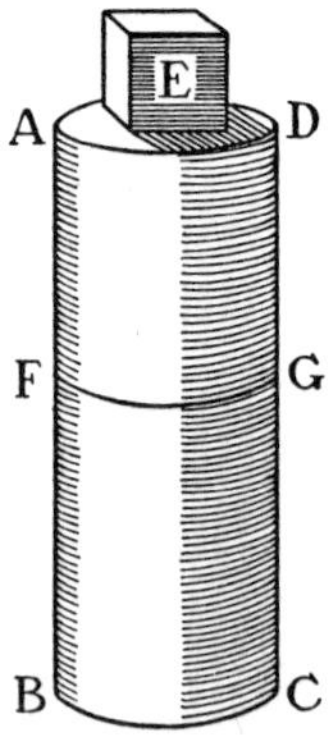

The vessel ABCD is a cylinder of wool or of some other compressible material (let us say air) and the vessel has two bases: BC stable, and AD movable and tight fitting. Let AD be loaded down by a block of lead E, which weighs 10,000,000 pounds. I think you will understand how much pressure the bottom BC is going to bear. Now if we force through FG a plane or cutting iron so that it enters and divides the compressed wool, I say that if the wool FBCG remains compressed as before, then even though the bottom BC does not any longer bear the weight placed on it by lead E, it will still be under strain just as it was before. Try it for yourself, for I shall not continue to bore you. Now for the second objection.

There was once a philosopher who, seeing his old servant put a faucet on a barrel, scoffed at him, saying that the wine would never come out because the nature of weights is to press downwards and not horizontally from the sides; but the servant made him see with his own eyes that although by nature liquids gravitate downwards, they press and spout in every direction, even upwards, as long as they find places to reach,—that is, places which resist with less force than their own. Sink a pitcher entirely in water with its mouth downwards, then make a hole through the bottom so that the air can come out, and you will see with what impetus the water moves up from below to fill it. You try it yourself, for I shall not annoy you further.

The third objection seems to me not quite apropos; it is certainly less valid than the others, even though it seems the strongest of all, since it is taken from geometry. That a body placed in water is counterpoised only by the weight of its own bulk of water is true; but it seems to me that the metal sustained in that neck of the vessel[1] can not be said to be immersed either in water, or in air, or in glass, or in a vacuum. We can only say that it is a fluid libratory body, one surface of which is bounded by a vacuum or near-vacuum which does not weigh anything at all, the other surface being bounded by air which is compressed by many miles of amassed air; and therefore that the surface which is not pressed at all rises because it is impelled by that other, and ascends until the weight of the metal raised comes to equalize the weight of the air pressing on the other side. Imagine the vessel A with a tube BCD joined to it and open at D, as represented; and let the vessel A be full of quicksilver.[2] It is certain that the metal will rise in the tube up to its level E; but if I immerse this instrument in water up to the mark F, the quicksilver will not rise to F, but

[1] The barometer tube.

[2] It is necessary (see below) that the top of A be open, or closed with flexible material; its other surfaces, like the tube, being rigid.

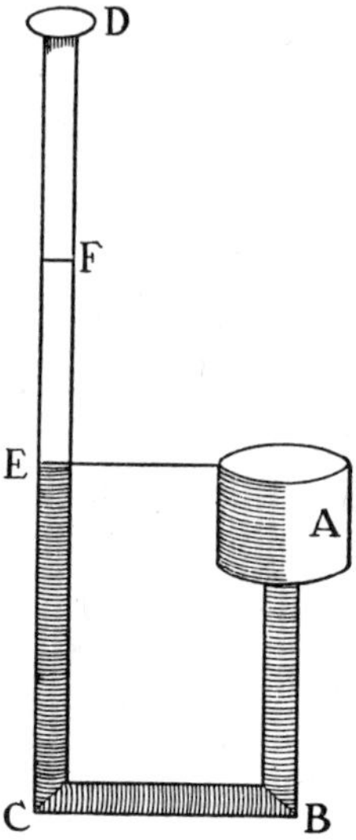

only until the height of the level in the tube will lie above the level of the vessel, EA, by about the fourteenth part of the height at which the water-[level] F will be above the level of the vessel A;[3] and this you may consider as certain as if you had performed the experiment yourself. Now here we notice that there can be a case in which the water is fourteen cubits high and the metal in the tube [section] ED is just one cubit high; then that cubit of metal alone does not counterbalance so much water, but only the water which is between A and F; and in these matters you know that one does not consider the widths and thicknesses of the solid bodies, but only vertical heights; and not absolute weights, but specific gravity.

But perhaps I have said too much; if I could speak to you personally, you would be better satisfied. I assure you that if anything else comes into your mind, you will be able to resolve all difficulties yourself, for many have been considered here and all of them are cleared up. . . .

[3] It is evidently assumed here that the change of the mercury level in A is, on account of the large diameter of the vessel, negligible.

BIBLIOGRAPHICAL NOTE

THE collected works of Pascal are available in two editions. That of L. Brunschvicg, P. Boutroux, and F. Gazier, *Œuvres de Blaise Pascal, publiées suivant l'ordre chronologique* (14 vols., Paris, 1904-14), is the richest source of information, though it lacks coherence on account of the chronological arrangement. Its introductions are lucid, and its criticisms suggestive. The preferable reference for all usual purposes is F. Strowski, *Œuvres complètes de Pascal* (3 vols., Paris, 1921-31), which is systematically arranged and coherent. The scientific work of Pascal, essentially though not fully complete, is contained in the first volume.

For biographical data and general criticism the reader may turn to E. Boutroux, *Pascal* (7th ed., Paris, 1919; English translation, Manchester, 1902), L. Brunschvicg, *Le Génie de Pascal* (Paris, 1924), or F. Strowski, *Pascal et son temps* (3 vols., Paris, 1921). All three of these are authoritative books by the editors of the standard editions of Pascal's collected works. Worth consulting also are Viscount St. Cyres, *Pascal* (New York, 1910), and G. Michaut, *Les Époques de la pensée de Pascal* (2d ed., Paris, 1902). The former is a comprehensive and satisfying narrative in English; the latter is a brief and interesting account, especially of the religious influences that affected Pascal and of his religious life.

In the field of scientific criticism August Heller, *Geschichte der Physik* (Stuttgart, 1884), II, 149-61, and F. Rosenberger, *Geschichte der Physik* (Braunschweig, 1882-90), II, 127-30, offer the best short accounts of Pascal's most important work and are useful as presenting its complete scientific setting.

I. Leavenworth, *The Physics of Pascal* (New York, 1930) is a critical narrative of the development of Pascal's work in natural science, with reference to that of his contemporaries, providing everything essential to a thorough general understanding. A comprehensive survey which presents in a critical manner, and adequately with respect to essentials, the whole historical background of Pascal's work in hydrostatics from the time of Aristotle is C. Thurot, "Recherches sur le principe d'Archimedes," published in *Revue archéologique* (N.S., XVIII [1868], 389-406; XIX [1869], 42-49, 111-23, 284-99, 345-60; XX [1869], 14-23). Cf. also T. L. Heath, *The Works of Archimedes* (Cambridge, 1897), P. Duhem, *Les Origines de la statique* (2 vols., Paris, 1905-6), and for the later periods relevant parts of the following: C. Thurot, "Les Expériences de Pascal sur le vide et la pesanteur de l'air" (*Journal de physique,* ser. 1, I [1872], 267-71), which consists in a short note on opinion concerning *horror vacui* and atmospheric pressure among Pascal's contemporaries; "Note historique sur l'expérience de Torricelli" (*ibid.*, pp. 171-76), by the same author, containing a French translation of Torricelli's letters to Ricci, with remarks on Galileo's relevant observations; C. Adam, "Pascal et Descartes" (*Revue philosophique,* XXIV [1887], 612-24; XXV [1888], 65-90), which provides a detailed account of the flux of opinion among Pascal, Descartes, and their contemporaries concerning atmospheric pressure, especially with reference to the conflict of their several metaphysical and scientific predispositions, prior to the experiment of the Puy de Dôme; F. Duhem, "Le Principe de Pascal, essai historique" (*Revue générale des sciences,* XVI [1905], 599-610), a somewhat detailed analysis of the work of Pascal's predecessors in hydrostatics, demonstrating clearly the exclusively methodical character of his own contribution and emphasizing its scientific value; by the same author, "Le Père Marin Mersenne et la pesanteur de l'air" (*ibid.*, XVII [1906], 769-82, 809-17), the first article being an historical survey of attempts to measure the weight of air, from Cardan to Descartes and Mersenne, the second, of more restricted interest, citing evidence to prove that it was Mersenne who first had the idea of testing Torricelli's theory of atmospheric pressure by measuring barometric heights at different altitudes; and two articles which discuss at length the delicate question of Pascal's seemingly egoistic attitude and unfairness to his predecessors and contemporaries—F. Mathieu, "Pascal et l'expérience du Puy de Dôme" (*Revue de Paris*, April and May, 1906; March and April, 1907), and G. Milhaud, "Pascal et les expériences du vide" (*Revue scientifique*, ser. 5, VII [1907], 769-77).

For the remaining literature, consult A. Maire, *Bibliographie générale des œuvres de Blaise Pascal* (Paris, 1925-27), I, "Pascal savant." The critical preface by Duhem and the introduction by Maire are valuable. See particularly works on scientific clubs (pp. 37-38), on the experiment of the Puy de Dôme (pp. 174-82), in criticism of Pascal's physics (pp. 183-242), and on scientific biography (pp. 276-99).

INDEX